Adnane MOUTAOUAKKIL
Mohammed EL MZIBRI
Abdelghani IDDAR

Radioimmunoassay for animal production

Adnane **MOUTAOUAKKIL**
Mohammed **EL MZIBRI**
Abdelghani **IDDAR**

Radioimmunoassay for animal production

Example of progesterone dosage in Moroccan goats

ScienciaScripts

INTRODUCTION

In Morocco, the breeding of small ruminants (sheep and goats) accounts for a large share of agricultural GDP. This activity, which plays an important socio-economic role, concerns more than 65% of the rural population (MADRPM, 2004). The livestock population, 95% of which is made up of local breeds, is estimated at 22.61 million head, including 5.35 million goats (MADRPM, 2004). Large pastoral and sylvo-pastoral areas contribute to the feeding of this livestock, particularly in the mountainous areas in the north of the country.

Goat breeding is particularly prevalent on small farms in remote and/or isolated areas (Benlakhal, 2004). Thus, 83% of goats are kept on farms smaller than 5 ha, and 90% are located in mountainous, pre-Saharan and Saharan areas (Benlakhal, 2004). In these areas, the more plastic goat is the only species capable of making the most of rugged terrain, making it the most important source of animal protein (milk and meat) for rural populations.

In northern Morocco, goat farming is of particular socio-economic importance and plays a vital role for local rural populations. Estimated at 733,000 head, i.e. 45% of the region's ruminant livestock and 15% of the national goat herd (Chentouf *et al.*, 2004), goat farming contributes around 60% of the income earned by farmers in this region (Fares and Ghalim, 1982; Jout and Karimi, 2004).

Currently, the national average for goat breeding productivity is estimated at 38 l of milk and 12 kg of meat per livestock unit per year (MAPM, 2008). This productivity remains relatively low compared with the technical resources that can be deployed to improve it. This is why, in its regional agricultural plan for the development of northern Morocco, the Moroccan Ministry of Agriculture is planning to implement an ambitious program aimed at improving milk production per goat per year from the current 120 l to 300 l by 2030. Achieving such a bold objective will necessarily involve improving breeding management.

In this respect, the need for better control of reproduction to increase herd profitability has led breeders and researchers to take a closer look at the various biotechnological methods that can be used to this end (El Amiri *et al.*, 2003). Controlling the physiological state of domestic animals has thus emerged as a key element in improving performance and/or reproductive management. In most species, the expression of this state depends largely on internal factors, such as steroid hormone levels (Thimonier, 2000). In this context, analysis of peripheral levels of progesterone (a steroid hormone playing an essential role in maintaining pregnancy) offers a powerful tool for better assessing the physiological state of females in a herd (Thimonier, 2000).

Peripheral plasma or serum progesterone levels vary considerably depending on the physiological state of the females in a farm (Lemon and Thimonier, 1973). During the anovulatory anestrus period, they are generally below 0.5 ng/ml. In ovulatory females, progesterone levels alternate between low values during the peri-ovulatory period and high values for most of the luteal phase. In pregnant females, after an evolution comparable to that observed during the early part of the cycle, progesterone levels remain high throughout gestation, irrespective of whether subsequent production is of solely ovarian or ovarian and placental origin (Sousa *et al.*, 2004).

What's more, after comparing the various methods for diagnosing pregnancy, it was found that the progesterone assay was the preferred method for early pregnancy diagnosis (El Amiri *et al.*, 2003). Applicable as early as 21ème days after fertilization in goats, this test offers pregnancy diagnosis accuracy values of up to 90%, while non-pregnancy diagnosis values approach 100% (Thimonier, 2000).

Progesterone measurement is thus an excellent indicator of mating, gestation, parturition and, in the case of artificial insemination, prediction of implantation failure (Thimonier, 2000). In recent years, its application has benefited greatly from the extremely important repercussions of the discovery of the immune system and its applications, particularly in the field of diagnostics. Indeed, the "antigen-antibody" interaction, the basis of immunological specificity, is currently used not only as a model for the study of molecular interactions, but also and above all as a detection and quantification tool, particularly in immunological assays (Neuburger, 2006).

On a practical level, the development of an immunoassay system for progesterone presents difficulties linked to the physico-chemical characteristics of this hormone. However, with the help of certain chemical coupling and labeling techniques, numerous solutions have been proposed to overcome this problem (Kothari and Pillai, 1998; Basu *et al.*, 2006; Simersky *et al.*, 2007). These solutions aim to considerably facilitate the development of highly sensitive and highly specific progesterone immunoassay kits (RIA and ELISA).

This is the background to our study, which forms part of the goat breeding development program in northern Morocco. Its overall long-term objective is to help improve the income of goat breeders by promoting this profitable and sustainable economic activity. The study focuses in particular on the development of new biotechnological techniques to improve herd productivity.

Specifically, the aim of our study is to develop a local kit for radioimmunoassay of progesterone for use in goat reproductive management. To develop such an assay procedure, we followed the following main steps:

- Production and purification of polyclonal anti-progesterone antibodies.
- Preparation and validation of a progesterone standard range.
- Preparation of radioactive progesterone tracers.
- Assembly of the progesterone radioimmunoassay device.

BIBLIOGRAPHICAL STUDY

I - GENERAL INFORMATION ON GOAT FARMING

Goats are one of the oldest domesticated species (7000 BC), which means, among other things, that man has controlled their reproduction for a very long time. They are found almost everywhere in the world, and are an important resource in many countries. However, data on their sexual behavior are much more limited than for cattle or sheep (Fabre-Nys, 2000).

One of the reasons for this relative paucity of information is that these animals are both independent and familiar, capable of adapting to a variety of environmental conditions, and have not posed any major problems for breeders (Gordon, 1997). It should probably be added that their breeding has not been subject to the same degree of intensification as that of cattle or sheep, and has therefore required less scientific investment to increase production. However, the performance and management of goat farms can be improved, notably by reducing the variability of fertility (40 to 85% after artificial insemination) or by controlling the reproduction period.

Sexual behavior is of obvious interest from this point of view. Breeding performance depends on reproduction, which in turn depends on the animals' willingness and ability to engage in sexual behavior and fertilize at the right time. This is true even when breeding involves artificial insemination, *in vitro* fertilization or embryo transfer. To best control the expression of this behavior, we need to know the various factors likely to influence it.

1. Physiology of goat reproduction

It's at puberty that an animal, male or female, is ready to reproduce. The age at puberty varies according to various factors (breed, time and method of birth, growth, etc.). In goats, 5-6 months for the young buck and 6-7 months for the female goat are commonly accepted as the age of onset of sexual activity (Brice, 2003).

From puberty onwards, the reproductive organs become functional, with periodic and seasonal activity. Sexual activity manifests itself in heat or estrus, externalized by a particular behavior. The goat is nervous, becoming abnormally agitated, riding and accepting rides from other females. She bleats, wags her tail frequently and loses her appetite.

Heat lasts an average of 36 h (varying from 24 to 48 h). In the absence of fertilization, if the female is in the autumn sexual season (or in spring after light conditioning), she returns to heat on average 21 days later (average length of the goat's sexual cycle). It should be pointed out, however, that at the start of the sexual season and after a male effect, short cycles of 5-6 days can be observed, generally with low fertility.

This sexual activity is controlled by various glands (hypothalamus, pituitary, epiphysis) which, thanks to the hormones they secrete, act on each other (by interaction) and on organs such as the ovary, uterus and udder (Figure 1).

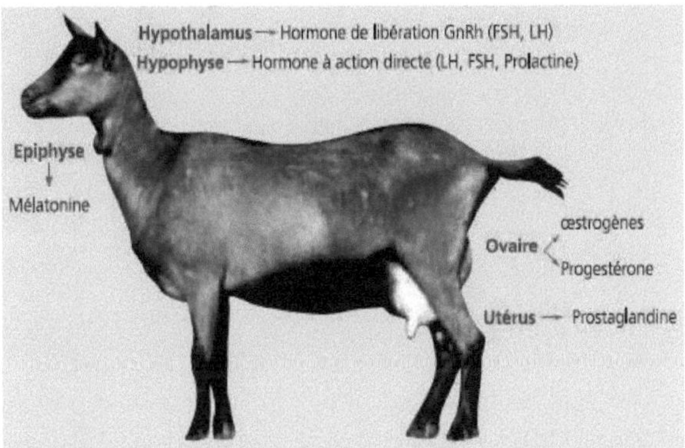

Figure 1: *Secretion of reproductive hormones in goats (Brice, 2003)*

Numerous external factors (climate, photoperiodism, season, diet, presence or absence of mates, etc.) also influence the functioning of these various glands. These hormonal balances and interactions with the external environment are also found in males.

2. Seasonality of goat reproduction

As with sheep, reproduction in goats has its own particularities due to seasonal sexual activity. The breeding season, controlled by the photoperiod, generally takes place during the waning days. Thus, most goat breeds bred in temperate and subtropical zones observe a period of sexual activity during autumn and winter (Figure 2), and a period of sexual rest during spring and summer (Valencia *et al.*, 1990).

Goats naturally come into heat from August to December (Figure 2). Farrowings take place from January to April, but most goats go into heat in January and February. In some months, however, there are heats without ovulation (especially at the beginning of the resumption of sexual activity) and ovulations without heat behavior (known as silent ovulations), mainly at the end of the sexual season. This phenomenon is found in almost all domestic ruminant species.

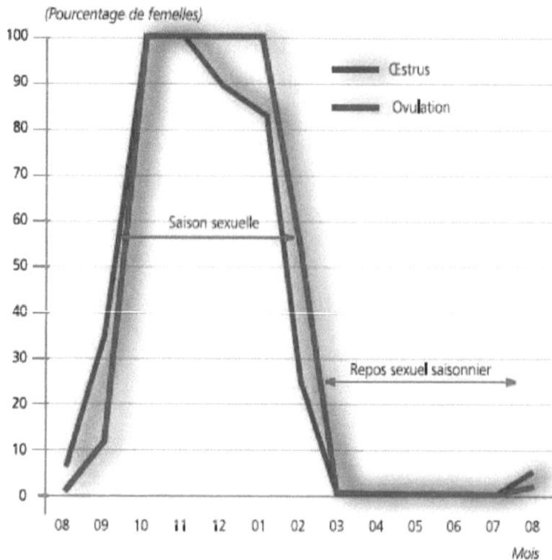

Figure 2: *Seasonal variations in the percentage of goats showing at least one estrus behavior or ovulation per month (Chemineau et al., 1992).*

After a more or less long period of sexual rest, a very small proportion of the female population presents a first estrus as early as June. It is often fertile, but in the absence of mating, it may be the first of a succession of estrus appearing at regular intervals (21 days), or it may be followed by a new period of anestrus. Cyclicity doesn't really kick in until September, before breaking off again in mid-December.

In billy goats, there are also seasonal variations in sexual activity, which peaks in autumn (Brice, 2003). Sperm production is highest from September to February, with testicular size increasing from October to November.

3. Endocrinology and reproductive control in goats

In most cases, estrus behavior occurs in goats with active ovaries, and is temporally linked to the females' ability to be fertilized (Figure 2). It is therefore logical to assume that the same signals involved in triggering estrus are also involved in ovulation (Fabre-Nys, 2000).

In both goats and ewes, estrus during the sexual season is preceded by a long luteal phase during which progesterone levels are high (4 to 12 ng/ml depending on the author Gonzalez *et al.*, 1992; Sawada *et al.*, 1995; Freitas *et al.*, 1997). Then, as progesterone levels fall following luteolysis, estradiol levels rise, as do androgen levels (Chemineau *et al.*, 1982; Homeida and Cooke, 1984; Mori and Kano, 1984). These increases are followed two days later by the onset of estrus behavior and the pre-ovulatory peak of the luteinizing hormone LH (Figure 3).

6

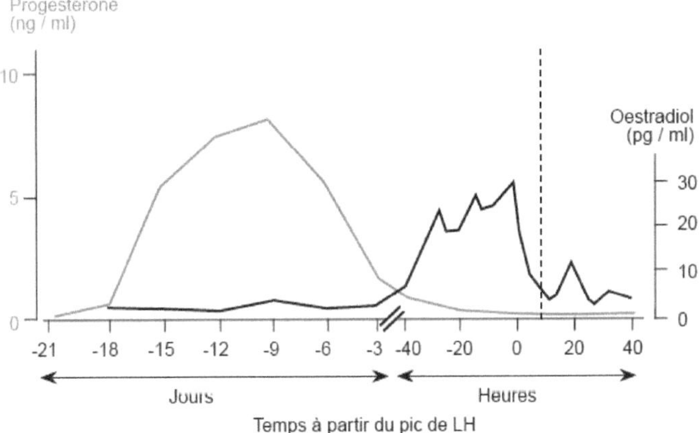

Figure 3: *Endocrine changes during the estrous cycle in goats*
(Okada et al., 1996)

Estradiol levels remain high for the first half of estrus, then fall sharply after the LH peak. In contrast to the ewe, estrus behavior in the goat also occurs early in the season, i.e. without a prior luteal phase, or after short cycles, i.e. after little or no progesterone secretion. The temporal evolution of these different changes suggests a stimulating effect of estrogens, and possibly androgens, and a modulating effect of progesterone.

Indeed, active estrone immunization blocks the effect of estrogen, increases progesterone levels and prevents the onset of estrus behavior (Chandrasekhar and Madan, 1996). Unlike in ewes, it is possible to induce all estrus behaviors (proceptivity and receptivity) in ovariectomized goats by administering estradiol alone (Sutherland and Lindsay, 1991; Kaplan and Katz, 1994). The minimum dose to induce estrus in 50% of females is 15 µg of estradiol benzoate. In this case, estrus appears with a latency of 25.5 h and lasts 12 h.

As in sheep, the estradiol sensitivity threshold varies with the season. It is minimal during the sexual season, i.e. on short days (Kaplan and Katz, 1994). The latency to estrus onset decreases and its duration increases with increasing doses (Sutherland and Lindsay, 1991; Billings and Katz, 1998). However, as in the ewe (Fabre-Nys *et al.*, 1993), the duration of estrus seems to depend essentially on the duration of estradiol presence (Okada *et al.*, 1998).

In ewes, progesterone has both a facilitating effect, when present prior to estradiol administration, and an inhibitory effect if present at the same time as estradiol. Both effects also appear to exist in goats, but the facilitating effect of progesterone is less clear.

Sutherland and Lindsay (1991) have shown that the administration of progesterone for 6 days prior to estradiol treatment does not alter the response rate as it does in the ewe, but only advances the onset of estrus, irrespective of the estradiol dose. For these authors, the role of progesterone is to increase estradiol sensitivity. Furthermore, Billings and Katz (1997) showed that the facilitating effect of progesterone varied according to the observed behavioral parameters and the season. The effect is particularly marked for attractiveness and receptivity during the anestrus period.

In contrast, the inhibitory effect of progesterone is clear in goats, as it is in sheep and pigs. Administration of progesterone during the 12 to 24 h preceding estradiol administration prevents the latter's effect (Billings and Katz, 1997). This inhibitory effect also affects LH secretion and ovulation. It is used in many treatments designed to control and synchronize the onset of estrus and ovulation (Baril *et al.*, 1993; Gordon, 1997).

4. Monitoring and diagnosing pregnancy in goats

Initiation of gestation in the goat begins with maintenance of the corpus luteum, preventing luteal regression induced during cycles by prostaglandin $F_{2\alpha}$, released by the endometrium. Gestation is then maintained by sustained progesterone levels, whose source is almost exclusively ovarian.

Gestation in goats varies from 146 to 155 days, depending on breed and individual (Ruckebush *et al.*, 1991). It is shorter in smaller goats. The average gestation period for local goats in northern Morocco is 149 days (Chentouf, 2007). This is between that of the Pakistani Dwarf goat (145 days) and that of the Toggenburg goat (152 days). Gestation length also varies according to litter size. Multiple-litter goats have shorter gestation periods than single-litter goats (Sousa *et al.*, 1999).

Early pregnancy diagnosis is also of great economic importance in goat production. It enables unsuccessful breeding or artificial insemination to be detected as early as possible, infertility cases to be identified and, if necessary, farm losses to be minimized through appropriate reforms. For example, it enables decisions to be taken to dry off lactating females at the right time, and to ensure appropriate feeding of pregnant females in order to optimize birth weight, improve viability of the young and prevent pregnancy toxemia.

The main methods used to monitor pregnancy in goats can be divided into two categories (Sousa *et al.*, 2004): On the one hand, clinical methods, including radiography, abdominorectal palpitation and ultrasonography. Secondly, laboratory methods, including hormone assays (estrone sulfate and progesterone) and pregnancy-associated protein (PAG) assays.

Among these methods, progesterone assay is of particular importance, both in the early diagnosis of management and in its monitoring and control. In fact, it is an excellent tool for the development of surveys or investigations into early or late embryonic mortality. It can also be used to develop a new preventive treatment (synchronization, super-ovulation) or to extend such a treatment to a breed or species whose pre-ovulatory physiology is poorly understood (Horie *et al.*, 2007).

II - GOAT FARMING IN MOROCCO

1. Goat populations in Morocco

In Morocco, there are three goat populations (Chentouf, 2007):

- *The Northern goat*: With its multicolored coat, the Northern goat is the result of crossbreeding between local goat populations and their Andalusian neighbors (Murciana and Granadina). This origin confers a dairy aptitude that can be seen in the size of the udder and the fineness of the skin. This dairy aptitude has been acquired through a selection process that breeders carry out within their herds, keeping the best dairy cows generation after generation. The annual milk production of this breed has been estimated by several authors, ranging from 59 to 96 l for lactation durations of 75 to 130 days (Hassani, 1997).

- *The Black goat*: Small in size and often black in color, this goat population is found in the Atlas mountains, on the plateaus and plains of the eastern and central regions, and in the pre-Sahara and Sahara. The milk production potential of this breed is estimated at 68 l per 90 days of lactation. The milk produced is highly suitable for cheese-making, with fat content of 68 g/l and protein content of 63 g/l (El Fadil, 1994).

- *The Drâa goat*: Medium-sized, with fine, supple skin and short hair. The coat varies in color (brown, piebald-black, etc.). This goat takes its name from the Drâa Valley, cradle of the breed and where it is mainly farmed. It is kept in small herds, in permanent stalls, and benefits from the by-products of oasis agriculture. This breed is characterized by the absence of a sexual season. Females give birth all year round. Its milk potential is higher than the national average: 153 l for a lactation period of 150 days (Hachi, 1990).

2. Goat production in Morocco

The national goat herd currently numbers 5.35 million animals, compared with an estimated 8 million in 1970 (MADRPM, 2004). This drop in numbers is due in part to the successive droughts that Morocco experienced in the 1980s, but also to the shift away from goat farming, considered as a subsidiary activity, towards other, more lucrative livestock.

Morocco's goat-breeding sector produces 20,000 tonnes of meat and 30 million liters of milk annually, representing barely 4% of national production for each of these two speculations (MADRPM, 2004). This situation is the result of the absence of an integrated government policy for the development of goat breeding.

However, the lack of interest shown by public authorities in goat farming is not in keeping with the vital role it plays on small farms in marginalized areas. Indeed, 90% of goat numbers are found in isolated mountain areas, with 40% in the Haut-Atlas, 25% in the Rif, 20% in the Moyen-Atlas and 5% in the Anti-Atlas (Benlakhal, 2004). In these areas, goat farming plays a very important socio-economic role for local rural populations. Thus, the intensification and development of goat farming is, in one way or another, a way of improving the living conditions of goat breeders in these regions.

This socio-economic importance, combined with the milk-producing capacity of local goats and the know-how of the local rural population in the production and valorization of goat's milk, motivated Moroccan public authorities to select milk production as a strategic option for the development of goat breeding in northern Morocco. In this respect, the objective set for goat farming is to increase red meat and milk productivity by 22%.

III - PROGESTERONE: ITS PHYSIOLOGICAL ROLE AND THE BENEFITS OF MEASURING IT

Progesterone is a steroid hormone synthesized from pregnenolone (a cholesterol derivative), mainly by the ovary and, to a lesser degree, by the adrenal glands, placenta and testis. It is a progestational hormone whose biological role is to promote implantation, then the development of gestation. It is distributed throughout the body in the same way as estrogen, but its biological effects are much more limited, particularly at the metabolic level (Moulin and Coquerel, 2002).

This hormone was first isolated from sow ovaries in 1932. Its chemical synthesis, however, was not made possible until 1934, thanks to the work of Butenandt (Simmer, 1975). Its chemical structure reveals a steroid with 21 carbon atoms and a molecular weight of 314.46 Da (Figure 4).

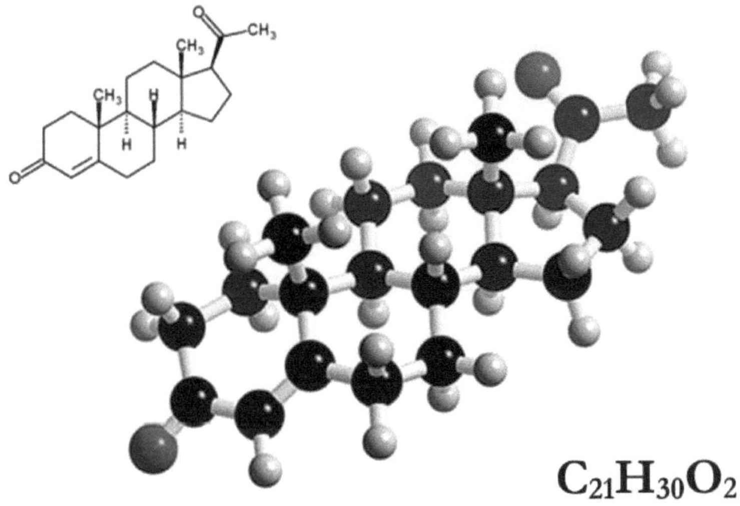

Figure 4: *Chemical structure of progesterone (4-pregnene-3,20-dione)*

1. Progesterone biosynthesis

The common precursor of all steroids is cholesterol, which is first converted to pregnenolone in the mitochondria by desmolase, then to progesterone by dehydrogenase and isomerase (Figure 5).

Progesterone itself is a metabolic intermediate that can lead to glucocorticosteroids (cortisol and corticosterone) synthesized in the fasciculated zone of the adrenal cortical glands, mineralocorticosteroids (aldosterone and deoxycorticosterone) synthesized in the glomerulated zone of these glands, and finally androgens (androstenedione and testosterone) synthesized in the reticulated zone.

Estrogens (estrone, estradiol) are obtained from androgens in numerous organs.

On the other hand, progesterone catabolism takes place mainly in the liver, where it is converted by several enzymes into pregnanedione, pregnanolone and finally pregnanediol.

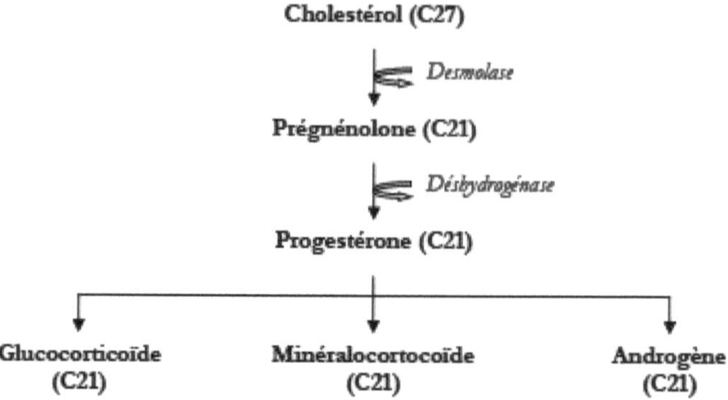

Figure 5: *Simplified diagram of steroid hormone biosynthesis (Argemi, 1998)*

2. Transport of progesterone

In plasma, progesterone is largely bound to transport proteins of high affinity but low specificity. In mammals, for example, progesterone and glucocorticosteroids are 90% bound to CBG (*cortisol binding globulin*, Mickelson *et al.*, 1981), while SBP (*sex steroid binding plasma protein*) binds testosterone and estradiol (Mercier-Bodard *et al.*, 1970). The free fraction, in constant equilibrium with the bound fraction, is solely responsible for activity. The precise role of these proteins remains to be elucidated. They could act as a reservoir in the event of a sudden drop in circulating steroid concentration, or facilitate transfer of the hormone from the producing cell to the circulation.

In tissues, progesterone (a lipophilic molecule) binds to fats, from which it can be progressively released. Its half-life in the body is estimated at around 30 min.

3. Mode of action of progesterone

Progesterone affects many cellular processes. Like all lipophilic hormones, it diffuses easily through the membranes of target cells. Once inside the cell, it binds very specifically and with high affinity (10 to 0.1 nmol) to a protein present in the nucleus in very small quantities: its receptor. The binding of progesterone induces a conformational change in the receptor, stabilizing its activated form. The receptor then has a strong affinity for a particular site on the chromatin: the acceptor. It is the binding of the receptor-progesterone complex to the acceptor that triggers gene expression and leads to the synthesis of the corresponding protein (Baulieu, 1992).

4. The physiological role of progesterone

As we have seen, progesterone acts by increasing the biosynthesis of several proteins. The resulting effects can be classified as follows:

4.1- Central effects

a- Regulation of pituitary secretion

The hypothalamus, through the secretion of LH-RH (*Luteinising Hormone - Releasing Hormone*), controls the episodic release of gonadotropic hormones (LH and FSH) into the general circulation. However, LH is not secreted continuously by the pituitary gland, but in pulses, defined by their frequency and amplitude, which stimulate the release of progesterone from the ovary in females (Olivereau, 1970).

However, progesterone plays a regulatory role in gonadotropin secretion, via hypothalamic LH-RH. It inhibits the secretion of pituitary gonadostimulins, or rather reduces the frequency of their secretion peaks and increases their amplitudes. This action involves the hypothalamic thermal regulation centers in the pre-optic area.

In the female, during the cycle just before ovulation, LH is massively released into the bloodstream to form the pre-ovulatory peak. This peak is caused by positive feedback from ovarian estrogens. The alternation of negative and positive feedback from the gonads to the hypothalamic-pituitary axis is an essential phenomenon in the control of gonadotropic activity.

b- Androgenic and anti-androgenic action

Progesterone has very low androgenic and anti-androgenic activity, which can be demonstrated at very high doses. Administered in high doses to castrated male rats, it can increase prostate weight (androgenic action), but when administered at the same time as testosterone, it reduces the latter's effect.

c- Anti-estrogenic action

Progesterone acts by inducing estrogen sulfotransferase and 17 α-hydroxy dehydrogenase, which accelerates the catabolism of estradiol to estrone.

4.2- Peripheral effects

There is a crucial prerequisite for progesterone's main action at nuclear receptors: the existence of estrogen, while the reverse is not true. This is the physiological sequence of the menstrual cycle.

In fact, progesterone alone has no precise influence on the target tissue at rest. It only acts on tissues already under estrogenic influence; indeed, only estrogens are capable of determining the synthesis and increase of progesterone receptor sites. Under these conditions, progesterone asserts both anti-estrogenic and tissue-specific properties (Zerr-Fouineau, 2006).

4.3- Progestin effects

✓ *In the uterus*: The essential function of progesterone is to prepare the uterus for implantation. It inhibits rhythmic contractions of the uterine musculature, creating a "uterine silence" without which gestation would be impossible.

✓ *In the endometrium*: Progesterone causes estrogen-induced mitosis to cease, giving rise to a secretory appearance known as "uterine lace", with glycogen-filled vacuoles. The endometrium is by far the richest tissue in progesterone receptors.

✓ *In the myometrium* : The tubal musculature is also significantly affected by the depressive influence of progesterone, which can modify tubal transport of the fertilized egg.

✓ *Cervix*: Progesterone suppresses estrogen-induced cervical mucus (glycoprotein secretion).

✓ *In the fallopian tubes*: Progesterone may slow egg transit.

✓ *In the mammary glands*: progesterone only acts on breast tissue prepared by estrogen. In synergy with estrogen, it then causes alveolar-acinar proliferation (Zerr-Fouineau, 2006).

4.4- Metabolic effects

The metabolic effects of progesterone are not very pronounced, if they exist at all. Only its natriuretic and diuretic action, of the anti-aldosterone type, is well established: progesterone competes with the mineralocorticoid at its receptors on the distal tubule.

a- Anti-mineralocorticoid effect

Progesterone inhibits the effect of aldosterone, stimulating sodium transport in renal cells, thereby reducing its plasma concentration and increasing urinary excretion (Champigny *et al.*, 1994).

b- Hyperthermizing effect

Progesterone is responsible for increasing temperature by around 0.5°C during the second half of the menstrual cycle in most mammals. Similarly, at plasma concentrations above 3 ng/ml, progesterone has a hyperthermia-inducing effect, raising basal body temperature by around 0.3 to 0.5°C.

4.5- Other effects

Progesterone also exerts an anesthetic effect and influences respiratory function through direct action on the corresponding centers. It may also have an effect on general thymia, probably via its cerebral and hypothalamic receptors. Finally, progesterone inhibits or reduces the increase in capillary permeability caused by estradiol (Pocock and Christopher, 2004).

5. The value of progesterone measurement for monitoring pregnancy in goats

The indispensable role of progesterone in maintaining gestation has long been known. It has been the basis for the development of hormonal diagnostic methods since the 1970s.

Minimal during estrus (0.2 to 0.6 ng/ml), progesterone concentration rises progressively from day 3 -4^{èmeème} (Figure 6), reaching a maximum of around 6 ng/ml between days 7ème and 10ème of the cycle in cycled (non-pregnant) goats. This concentration remains stable until around 16 -17^{èmeème} days, after which it falls sharply as a result of luteolysis induced by uterine prostaglandin $F_{2\alpha}$. In the event of fertilization, progesterone levels are maintained by tau interferon, the embryonic signal in ruminants (Sousa *et al.*, 2004). It should be noted that progesterone secretion in goats is almost exclusively of ovarian origin.

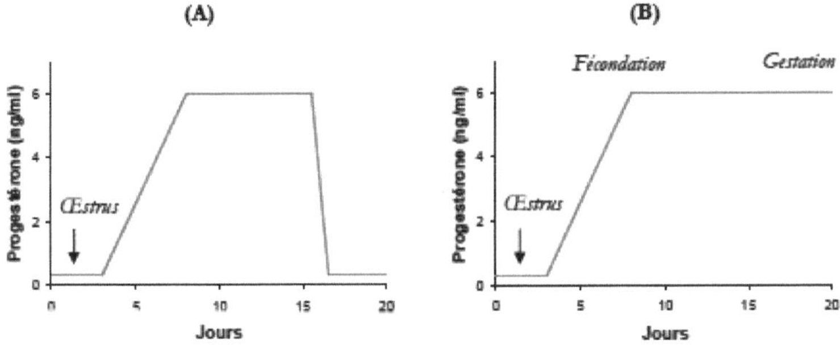

Figure 6: *Theoretical evolution of plasma progesterone concentration in goats during the sexual cycle (A) and gestation (B) (Sousa et al., 2004).*

The main advantage of progesterone measurement is that it enables early diagnosis as early as 21 -22^{èmeème} days in goats (Thimonier, 2000). However, this diagnosis is not without its faults. While it can be considered indisputable when negative, enabling early identification of non-pregnant animals, it only demonstrates the presence of a functional corpus luteum when positive (> 2 ng/ml). It can give rise to a number of false positive diagnoses, depending on the extent of embryonic mortality in the herd, or in the case of imprecise knowledge of the date of fertilization.

In fact, studies on this subject report that up to 10-30% of goats classified as positive are not in fact positive (Vanroose *et al.*, 2000). In these animals, elevated progesterone levels are indicative of an altered cycle (shorter or longer cycle), the presence of a luteal cyst, genital tract infections or the occurrence of early or late embryonic mortalities, which may induce persistent corpus luteum or pseudopregnancy in the goat.

We can conclude that this method is effective in diagnosing a non-pregnant state, not only enabling animals to be returned to reproduction without delay, but also ensuring treatment in the event of utero-ovarian pathologies.

6. Progesterone dosage methods

For all the above reasons, clinical analysis of progesterone levels has been in use for almost 30 years. Early tests were based on chemical reactions. These methods were highly specific, but required large quantities of the steroid (in the milligram range). The development of immunoassays has made it possible to obtain assays that are much more sensitive (in the hundreds of pg range) and easier to perform.

However, the major problem with these assays concerns the specificity of the antibodies used. This problem is all the greater when it comes to the specific recognition of a small, non-immunogenic molecule (hapten), which is hydrophobic and above all possesses numerous analogues with very similar structures, as is the case with progesterone (structural similarity with testosterone, cortisol, estradiol, pregnenolone, etc.).

To reduce cross-reactivity, one solution was to extract the steroids using organic solvents and separate them by chromatography before assaying them by immunoanalysis (Barkley *et al.*, 1985; Agasan *et al.*, 1994; Wudt *et al.*, 1995). However, this method is very cumbersome and unsuitable for intensive, routine use.

Highly specific antibodies remain the key to immunoassays. Methods for obtaining them have proliferated in recent years, providing a means of obtaining highly accurate assays without the need for expensive or difficult-to-use routine equipment. The precision of these assays relies on the ability of antibodies to recognize only their antigen. This has become possible thanks to the use of immunogens prepared by grafting the steroid onto a carrier protein at certain positions on the steroid core (Kothari and Pillai, 1998; Basu *et al.*, 2006; Samuel *et al.*, 2006).

This progress made it possible to obtain highly specific polyclonal anti-progesterone antibodies, which have been used in the development of numerous immunoassays for this hormone (Erlanger *et al.*, 1958; Bacigalupo *et al.*, 1987; Basu *et al.*, 2006). At present, these assays are performed by radioimmunoassay (RIA) or enzyme immunoassay (EIA). They can be performed on blood samples, whole milk, skimmed milk or milk cream. Recent studies have reported that it is even possible to measure progesterone in the saliva of certain animals.

In terms of recommendations, it should be remembered that for progesterone immunoassays, blood samples should be centrifuged quickly after collection, or collected in tubes containing sodium azide. Milk samples should also be preserved by adding sodium azide, potassium dichromate or mercuric chloride.

IV - IMMUNOASSAYS

1. The "antigen - antibody" interaction

1.1- Definition of antibodies and antigens

Antibodies (Ac) or immunoglobulins (Ig) are one of the main components of the immune system, helping to defend the body against the intrusion of pathogens. The majority of circulating (blood-borne) antibodies are produced by plasma cells, corresponding to differentiated B lymphocytes, in response to the penetration of an exogenous substance into the body. This production of antibodies against an "aggressor" is in fact only part of the immune response (humoral response). The introduction of an immunogenic substance into the body stimulates a large number of lymphocytes, which produce a large number of antibodies recognizing different areas of this substance called epitopes or antigenic determinants (polyclonal response).

Antigens are defined as any structure capable of being recognized by antibodies (antigenic potency). They can be extremely diverse in nature and characteristics (proteins, polyosides, nucleic acids, lipids, steroids, synthetic molecules), molecular mass and conformation. Some are capable of eliciting an immune response in the form of antibody synthesis (immunogenicity), depending of course on the immunological potential of the host (Van Regenmortel, 1994). Other antigens, known as haptens, are small natural or synthetic molecules (MW < 5000 Da), with antigenic reactivity but no immunogenic power, except after coupling to a *carrier* molecule. This is the case with steroids, including progesterone.

1.2- Origins of immunoanalysis

The main stages in the development of immunoanalysis are summarized in Table 1. The discovery of the humoral support of immunity dates back to the end of the 19th[ème] century, with the work of Emil Von Behring in 1890 on diphtheria antitoxin (Nobel Prize 1901), followed by that of Paul Ehrlich (Nobel Prize 1908), which laid the foundations of immunochemistry.

The discovery of the main phenomena and the invention of precipitation techniques (by Kraus) and synthetic haptens (by Letsteiner) led to the formulation of fundamental theories concerning the "antibody-antigen" reaction. Arrhenius introduced the idea of reversibility, i.e. equilibrium; Heidelberger showed that the valence of antibodies was two (also confirmed by Karush, 1958); Letsteiner and Pauling initiated the thermodynamic characterization of "antigen - antibody" reactions and highlighted the phenomenon of "cross-reactivity".

The use of antigenic immunoglobulin antibodies to classify immunoglobulins (isotypy, allotypy, idiotypy).

Dissection of the antibodies was undertaken thanks to the complementary work of Porter and Edelman (Porter, 1967; Edelman *et al.*, 1969). The three-dimensional structure of the antibodies and their binding sites was clarified by X-ray crystallographic studies (Poljak *et al.*, 1973).

The nature of antibodies and their binding to antigens has become a powerful tool for exploration and quantification. The development of radioimmunoassay is classically attributed to Yalow and Berson, who in 1959 developed the first immunoassay for human insulin in plasma (Yalow and Berson, 1960).

The discovery of new assay principles, both competitive and immunometric (Wide *et al.*, 1967; Miles and Hales, 1968; Ling and Overby, 1972), the evolution of techniques for separating free or bound species in the immune complex (Catt *et al.*, 1967), and the search for original means of revelation, such as enzyme labelling (Engvall and Perlman, 1971), gave these methods a new lease of life.

In 1975, the work of Köhler and Milstein made it possible to obtain monoclonal antibodies (mAbs), produced by a single lymphocyte clone (Köhler and Milstein, 1975). In addition to its scientific interest, this discovery had major technological repercussions in the industrial use of immunological specificity, thanks to the possibility of obtaining large quantities of perfectly homogeneous monoclonal antibodies with a given specificity, which could be produced at will.

Table 1: *Stages in the development of immunoanalysis*

Year	*Author*	*Method discovery or development*
1890	Von Behring	Antitoxins
1896	Gruber and Durham	Agglutination
1892	Fisher	The "lock *and key*" model
1897	Ehrlich	The chemical basis of antitoxic specificity *Side-chain* theory
1897	Kraus	The rush
1917	Letsteiner	Synthetic haptens
1929	Heidelberger	Quantitative chemical serology

1937	Langmuir and Schaefer	Solid-phase immunoassay (first description)
1938	Kabat	Antibodies are gamma globulins
1942	Coons	Immunofluorescence
1946	Oudin and Ouchterlöny	Immunodiffusion
1953	Grabar	Immunoelectrophoresis
1958	Porter	The structure of immunoglobulins
1959	Edelman	The sequence of an immunoglobulin
1959	Yalow and Berson	Radioimmunoassays
1966, 1967	Wide and Porath, Catt and Tregear	The use of antibodies covalently bound to a solid phase
1967, 1968	Wide *et al.*, Miles and Hales	Immunoradiometric assay
1971	Engvall and Perlman	Enzyme immunoassay
1972	Ling and Overby	Two-site immunoradiometric assay (sandwich assay)
1973	Poljak *et al.*	The three-dimensional structure of an antibody using crystallography
1975	Köhler and Milstein	Monoclonal antibody production

Based on E. Zuber's thesis (Zuber, 1997).

1.3- Antibody structure

Antibodies are glycoproteins with a common basic structure, generally consisting of two light polypeptide chains (or L for *Light*) with a molecular weight of 25 kDa and two heavy chains (or H for *Heavy*) with a molecular weight of between 50 and 72 kDa. It should be noted that some immunoglobulins (Ig) can associate in multimeric structures.

IgGs are the main antibodies used in immunoanalysis. For a given antibody, the two heavy chains are identical, as are the two light chains. Each heavy chain is closely associated with one of the light chains; stability is ensured by intra- and inter-catenary disulfide bridges and by tertiary and quaternary interactions. A classic distinction is made between a constant (C) and a variable (V) part, consisting of the amino-terminal region of the heavy and light chains. This symmetrical structure features two antigen-binding sites (paratopes) in the variable part (divalence), as shown in Figure 7.

IgG digestion with papain yields three distinct proteolytic fragments: an Fc fragment (*crystalline fragment*) and two monovalent fragments, still possessing the ability to bind antigen, known as Fab (*fragment antigen-binding*). The paratopes are shared between the VH and VL modules, which combine to form a barrel-like structure. Sequence analysis of the variable parts defines 3 regions per module where primary sequence variability is high, known as hypervariable loops, or *CDRs* (*complementarity determining regions),* which enable antigen recognition. This interaction is increasingly well characterized both structurally and functionally (Van Regenmortel, 1998). The two essential characteristics of molecular recognition are affinity and specificity.

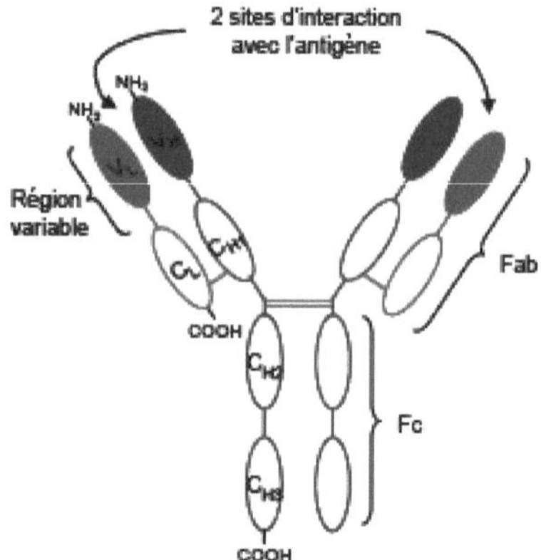

Figure 7: *Simplified structure of a basic immunoglobulin unit*
Heavy chain (index H) in blue, light chain (index L) in green,
V: variable part, C: constant part, Red lines symbolize disulfide bridges,
Fab: fragment antigen binding; Fc: fragment crystalline

1.4- Notions of affinity and specificity

The affinity of the interaction defines the stability of the Ag-Ac complex, and can be characterized quantitatively by the affinity constant K_a (equilibrium constant, in M^{-1}), written as a function of the reaction equilibrium.

In simplified terms, the kinetic reaction between a binding site (paratope) and an antigenic determinant (epitope) can be written as follows:

$$[1] \qquad Ac + Ag \xrightarrow[k_{off}]{k_{on}} Ac - Ag$$

with Ac: paratope; Ag: epitope; Ac-Ag: epitope-paratope complex; k_{on} : association rate constant; k_{off} : dissociation rate constant.

The concentration of [Ac-Ag] complex therefore evolves over time according to the following differential equation:

$$[2] \qquad \frac{d[Ac - Ag]}{dt} = k_{on}[Ac][Ag] - k_{off}[Ac - Ag]$$

with, in square brackets, concentrations (M), functions of time t (s); k_{on} (M s^{-1-1}); k_{off} (s).$^{-1}$

The affinity constant K_a of a binding site for an antigenic site is defined by the relation :

[3] $$K_a = \frac{k_{on}}{k_{off}}$$

At equilibrium, the concentration of [Ac-Ag] complex is stable, hence :

[4] $$K_a = \frac{[Ac - Ag]}{[Ac][Ag]}$$

Classically, K_a is evaluated by equilibrium measurement using the Scatchard representation (Yalow and Berson, 1960). It can also be determined using the Biacore biosensor to calculate the kinetic constants k_{on} and k_{off} (Fivash *et al.*, 1998). The greater the affinity of the Ag-Ac interaction, the higher the K_a and the more stable the complex. The antibody is generally considered to have good affinity for the antigen when K_a is greater than or equal to 10^9 M^{-1} .

In the case of a multivalent molecule, a cooperative effect between sites has been observed, giving rise to the notion of "avidity". The authors observed that the affinity of a multivalent molecule was clearly greater than the arithmetic sum of the affinities of each of its sites (Dmitriev *et al.*, 2003).

Antibody specificity is a much more abstract concept than affinity. It corresponds to the ability of antibodies to select between closely related chemical species (Janin, 1995). Ideally, a "specific" antibody would form stable complexes only with the antigen against which it was produced. However, there are always cross-reactions, with the antibody recognizing with varying affinities analogous antigens of more or less similar structure to that of the reference antigen. The specificity of an antibody can therefore only be relative, given the range of possible antigens. In a given biological medium, an antibody is considered to have good specificity when it can selectively measure the antigen among the molecules present in the sample (Volland, 1999). The affinity and specificity of the Ac-Ag interaction strongly influence the performance of immunoassays.

2. Immunoassays

2.1- General principle

Numerous analytical methods based on the use of antibodies have been described. In this section, we will only consider the case of quantitative assays, which represent the vast majority of assays performed for research or diagnostic purposes.

Immunoassays are relative analysis methods based on calibration curves, established by carrying out several equilibria with known and variable concentrations of the molecule to be assayed (analyte). Here, we'll confine ourselves to the case of measuring antigen concentration. By measuring the quantity of complexes formed at each equilibrium, we can establish an experimental relationship between the concentration of the antigen and that of the Ag-Ac complex. Measuring the concentration of the Ag-Ac complex is often facilitated by the use of labelled molecules or tracers.

Immunoassays can be classified according to different keys. A distinction can be made between methodologies enabling monitoring of a reaction carried out at a solid-liquid interface, and those involving a reaction in a homogeneous liquid medium. We can also distinguish between assays employing a single antibody, or two antibodies recognizing different epitopes ("sandwich" assays). They can also be classified according to the type of marker or detection method employed.

2.2- "Homogeneous format" and "heterogeneous format" dosages

Classically, there are two types of assay, depending on whether or not a separation step has been performed on the immunological complexes (so-called heterogeneous *versus* homogeneous assays).

Homogeneous format" assays are performed in a single medium. They are used when the formation of the Ag-Ac complex modifies the signal carried by the tracer: the formation of the Ag-Ac complex can then be monitored directly by measuring the "global" signal of   the   solution. "signal of the solution. In this way, the measurement is carried out on all the components of the immunological equilibrium, without the need for prior separation.

In contrast, so-called "heterogeneous format" assays involve a separation of the uncomplexed tracer from that involved in the Ag-Ac complexes, following the immunological reaction. In practice, this separation, which must be carried out before the final measurement, is often achieved using two media or phases: a liquid phase and a solid phase. Ac or Ag is bound to a solid phase (e.g. tube wall, microplate wall, magnetic beads) by simple adsorption or by various types of more specific binding. Heterogeneous assays, which generally involve interrupting the reaction to measure the signal, do not easily allow rapid kinetics to be followed, and are less amenable to automation. On the other hand, they do offer the advantage of being able to partially overcome the matrix effect by separating the uncomplexed fraction before measurement.

2.3- Competitive and immunometric formats

These assays can be performed in two formats: competitive and immunometric (Figure 8). Competitive assays are based on competition between a labeled antigen (or Ag*) and an unlabeled antigen (the analyte) for binding to Ac binding sites (paratope). This type of assay can be represented by the following two equilibria:

$$Ac + Ag \xrightarrow[k_{off}]{k_{on}} Ac - Ag \qquad \text{and} \qquad Ac + Ag* \xrightarrow[k*_{off}]{k*_{on}} Ac - Ag*$$

Ac and Ag* tracer are used at predetermined, constant concentrations. To allow competition and achieve good sensitivity, these reagents must be used in limited quantities. The concentration of unlabeled antigen is either known to establish the calibration curve, or unknown in the samples. The higher the Ag concentration, the more the tracer binding is displaced: the signal obtained is therefore generally inversely correlated with Ag concentration.

In immunometric assays (Figure 8), the tracer is a labeled antibody instead of a labeled antigen. The principle involves sandwiching the antigen to be assayed between two antibodies, which are used in excess of the Ag. Generally speaking, the signal increases with the concentration of antigen to be assayed (linearly for low concentrations). The major advantage of this assay format lies in the use of reagents in excess, favoring complex formation and resulting in greater sensitivity than competitive assays using the same antibodies, with a gain in sensitivity of 10 to 1000 (Ekins, 1993).

On the other hand, as this method usually involves the simultaneous binding of two antibody molecules to the analyte, it is less readily applicable to the assay of low molecular weight molecules. Indeed, this method requires two antibodies recognizing two different epitopes on the antigen (or complementary antibodies). However, several methods have been described for the immunometric determination of haptens (Pradelles *et al.*, 1994; Creminon *et al.*, 1995; Volland, 1999).

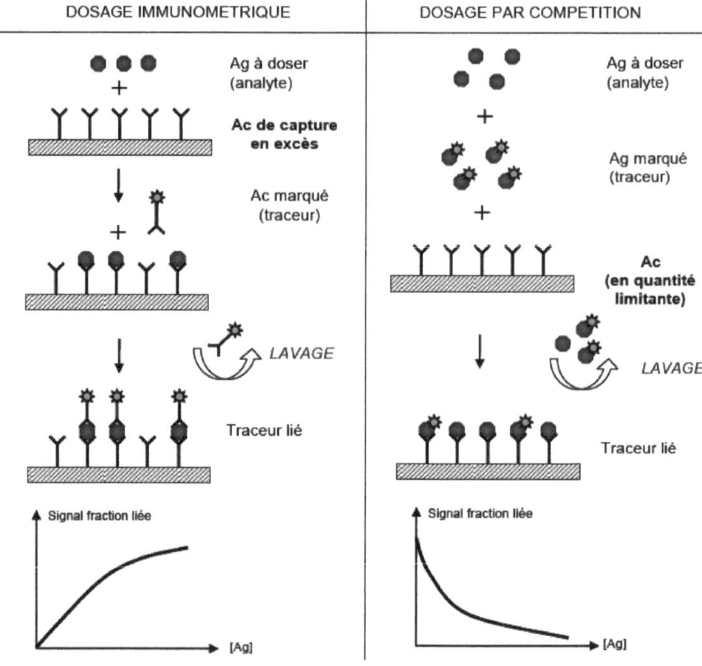

Figure 8: *Schematic diagram of immunometric and competitive assays (solid phase)*

In the immunometric format, the analyte is sandwiched between two antibodies.
In the competitive format, the analyte and the tracer compete for binding to the sites.
antibody binding.

2.4- Detection methods

Most immunoassays use labeled molecules to reveal the immune complex formed. This signal is carried by a marker linked (covalently or not) to the Ac or Ag. Unlabeled systems do exist, notably the Biacore biosensor, based on plasmon resonance measurement, but this is rarely used for routine assays. Numerous markers can be used (Pelizzola *et al.*, 1995). These are usually radioactive, fluorescent or luminescent markers, or enzymes (with the formation of a colored product).

The first assays developed used radioactive tracers (Yalow and Berson, 1960), notably Tritium (^{3} H) and Iodine 125 (^{125} I), the latter being the most widely used marker in radioimmunoassay, due to its high specific activity. RIA (*Radio ImmunoAssay*) assays are generally highly sensitive, enabling direct signal measurement. What's more, radioactive labelling (with^3 H in particular) causes very little change to the molecule, thus preserving its immunoreactivity. However, for safety reasons, they require very careful handling, and are rarely fully automated. Radioactive waste management is costly, and tracers need to be renewed frequently because of their radioactive decay (e.g. half-life of 60 days for[125] I). Today, radioisotopes account for only a small fraction of immunoassays (Brossier *et al.*, 2001).

Thanks to the work of Engvall and Perlman (Engvall and Perlman, 1971), numerous enzyme *immunoassay* (EIA) methods have been developed. The concentration of labeled antigen is determined using a substrate, generally modified from the natural substrate, which will react with the enzyme to form a signal-bearing product (usually colored, fluorescent or chemiluminescent). Enzyme-linked immunosorbent assays are used routinely for the determination of a wide range of compounds, and can achieve very high sensitivities (down to attomoles i.e. 10^{-18} moles, or even zeptomoles i.e. 10^{-21} moles of analyte), thanks in particular to the signal amplification inherent in enzyme *turnover* (Cousino *et al.*, 1997). The most commonly used enzymes are horseradish peroxidase, alkaline phosphatase and β-D-galactosidase. In general, EIA assays are used in a heterogeneous format with indirect signal measurement: after removal of the tracer not involved in the immunological complexes by a separation step (wash step), the substrate is added; then, after enzymatic reaction (revelation step), the product formed is quantified.

In recent years, new fluorescence immunoassay methods have been developed. These methods are distinguished by their high sensitivity (Harma *et al.*, 2000; Schultz *et al.*, 2003) and rapidity. They enable direct measurement, avoiding any revelation step. However, their applications are still limited and sometimes require costly adaptations.

EXPERIMENTAL STUDY

PART 1: PRODUCTION AND PURIFICATION OF POLYCLONAL ANTI-PROGESTERONE ANTIBODIES

Progesterone is a non-immunogenic molecule due to its low molecular weight (314.46 g/mol). It is therefore incapable, in its free state, of stimulating an organism's immune response to produce antibodies. Thus, the production of polyclonal antibodies directed against this hormone requires its prior coupling with a high-molecular-weight carrier protein.

Several progesterone-protein coupling techniques have been proposed (Erlanger *et al.*, 1958; Bacigalupo *et al.*, 1987; Basu *et al.*, 2006). Most of these techniques highlight the influence of the conjugation site (binding position of the protein to the progesterone) and the nature of the ligand (binding arm) on the affinity and specificity of the antibodies produced (Niswender, 1973). Indeed, it has been shown that progesterone-protein immunogens conjugated to position 3 or 11 of progesterone (Figure 9) produce antibodies with higher affinities and specificities than those developed against immunogens conjugated to other positions of the steroid (Kothari and Pillai, 1998; Samuel *et al.*, 2006).

It is for this reason that we proposed to prepare immunogens, conjugated to one of these two positions (3 and 11) of progesterone, in order to use them to locally produce highly specific polyclonal anti-progesterone antibodies, necessary for the development of immunoassay systems for this hormone.

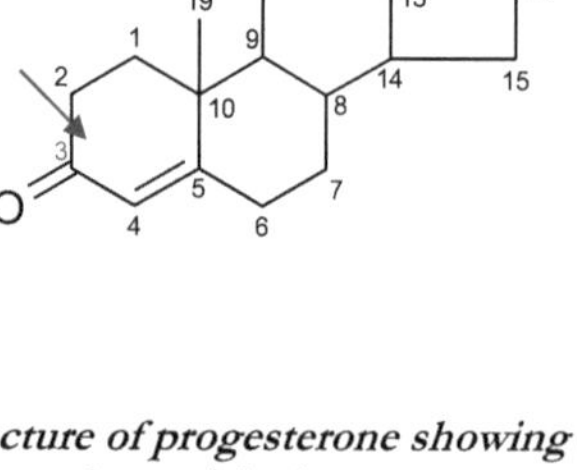

Figure 9. *Chemical structure of progesterone showing conjugation sites ( )→ most commonly used for immunogen preparation*

I - IMMUNOGEN PREPARATION

The progesterone derivatives generally used for the preparation of immunogens are: progesterone 11α-hemisuccinate (P11α-HS) and progesterone 3-(O-caroboxymethyl) oxime (P3-CMO). Both derivatives are characterized by the presence of a carboxyl group, which plays an essential role in coupling reactions (Figure 10).

In the absence of commercialization and/or excessively high prices, these derivatives have been synthesized in our laboratory using 11α-hydroxyprogesterone and progesterone as precursors, respectively for the synthesis of progesterone 11α-hemisuccinate and progesterone 3-(O-caroboxy-methyl) oxime.

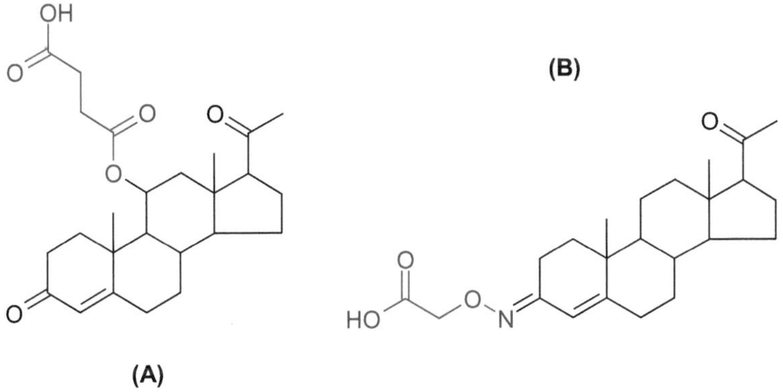

Figure 10. *Chemical structures of progesterone 11α-hemisuccinate* (A)
and progesterone 3-(O-caroboxymethyl) oxime (B)

These progesterone derivatives are most often coupled to bovine serum albumin (BSA), although other high-molecular weight proteins, such as hemocyanin (KLH) or ovalbumin (OVA), can also be used as carriers.

1. Synthesis of carboxyl-function progesterone derivatives

1-1. Progesterone 11α-hemisuccinate (P11α-HS)

Progesterone 11α-hemisuccinate (P11α-HS) was synthesized from 11α-hydroxyprogesterone (P11α-OH) using the method described by Allen and Redshaw (1978), which we adapted to our laboratory conditions.

a- Reaction scheme

11α-Hydroxyprogesterone

Progesterone 11α-hemisuccinate

b- Materials required

Reagents	Special equipment
- 11α-Hydroxyprogesterone	- Balloon heater
- Succinic anhydride	- Balloon
- Pyridine	- Refrigerant
- Ethyl acetate	- Rotavapeur
- HCl	- Decanting funnel
- KOH	
- Benzene	
- Hexane	
- $MgSO_4$	

c- Experimental protocol

Dissolve 250 mg (0.75 mmol) of 11α-hydroxyprogesterone (Sigma-Aldrich) and 250 mg (2.5 mmol) of succinic anhydride in 10 ml of pyridine. The reaction mixture was heated and left to reflux for 5-6 h. After removal of the solvent (pyridine) using the rotavapor (85°C, in vacuo), the resulting residue (semi-solid) was taken up in 50 ml ethyl acetate, then extracted with a 4% KOH solution (2 × 50 ml).

The recovered aqueous phase is washed with ethyl acetate, then acidified to pH 2 with concentrated HCl solution. The cooled mixture is then extracted with ethyl acetate (2 × 50 ml). The organic extract obtained is washed with distilled water (100 ml) before being dried over $MgSO_4$.

After filtration, the solvent is removed using a rotavapor (50°C, under vacuum) and the residue obtained is progesterone 11α-hemisuccinate.

Recrystallization of the reaction product (P11α-HS) is possible after hot solubilization of the recovered residue in a benzene/hexane mixture (v/v, 5 ml) and cooling (4°C).

d- Physicochemical analysis

The P11α-HS synthesis reaction was monitored by thin-layer chromatography using silica gel as the stationary phase and a chloroform/acetone mixture (7/3; v/v) as the mobile phase (eluent).

The results obtained after UV revelation (Figure 11) show the disappearance of the starting reagent and the appearance of a new product with an R_f of 0.16 corresponding to that of P11α-HS according to the literature (Allen and Redshaw, 1978).

Figure 11: *Silica gel TLC analysis of synthesized P11α-HS*
R *being the initial reactant;* P₁ *being the reaction product.*
The elution system used is a chloroform/acetone mixture (7/3; v/v).

The melting point of the synthesized product (P11α-HS) was evaluated at 154°C using the *"Electrothermal-9300 Digital Melting Points Apparatus"*. This result is in perfect agreement with that reported in the literature, which stipulates a melting point of this product between 152-154°C (Allen and Redshaw, 1978).

In order to better elucidate the chemical structure of the synthesized product (P11α-HS), a complete analytical study was carried out at the Unité d'Appui Technique à la Recherche Scientifique (UATRS), of the Centre National pour la Recherche Scientifique et Technique (CNRST). This study includes the following analyses: IR spectrometry, NMR[1] H and mass spectrometry.

In particular, the IR spectrum recorded for the synthesized P11α-HS shows a broad band at 3442 cm^{-1} corresponding to the valence vibrations v_{O-H} of the carboxyl function, and an intense band at 1729 cm^{-1} corresponding to the valence vibrations $v_{C=O}$ of the carbonyl acid function (Figure 12).

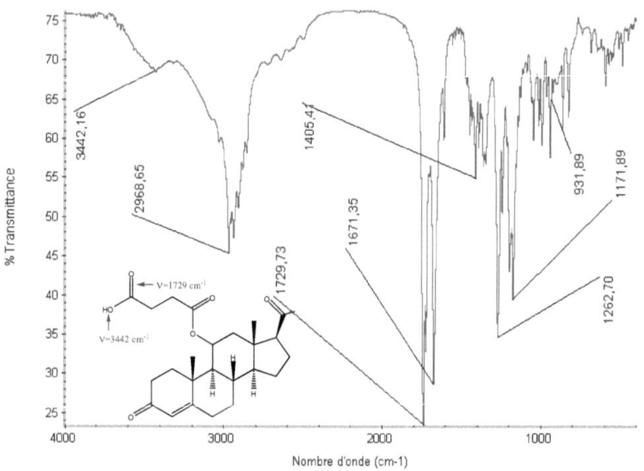

Figure 12: *IR spectrum of synthesized P11α-HS*
Solid sample; KBr

The[1] H NMR spectrum of the synthesized P11α-HS taken in deuterated chloroform shows, in addition to signals due to progesterone protons, two multiplets centered at 2.56 and 2.70 ppm respectively, attributed to protons of the hemisuccinate chain introduced on the starting molecule (Figure 13).

On the other hand, the mass spectrum of the synthesized P11α-HS in particular shows a molecular peak at m/z = 431.06 corresponding to [M + H] of this product (Figure 14).

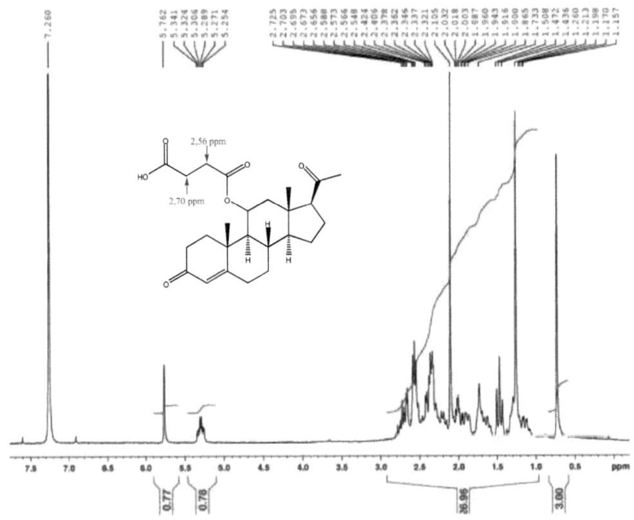

Figure 13: *^{1} H NMR spectrum of synthesized P11α-HS*
Sample taken in deuterated chloroform

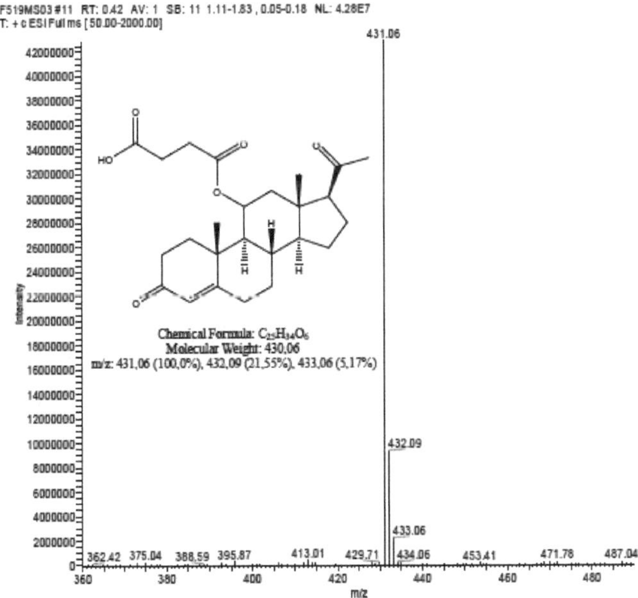

Figure 14: *Mass spectrum of synthesized P11α-HS*

31

Taken together, these results, obtained by means of IR, NMR[1] H and mass spectroscopy, clearly show that the product synthesized is progesterone 11α-hemisuccinate (P11α-HS). The yield of the synthesis reaction is estimated at 70%.

1-2. Progesterone 3-(O-caroboxy-methyl) oxime (P3-CMO)

Progesterone 3-(O-caroboxy-methyl) oxime (P3-CMO) was synthesized from progesterone using the method described by Janosky *et al.* (1973), which we adapted to our laboratory conditions.

a- Schéma réactionnel

b- Materials required

Reagents	Special equipment

- Progesterone	- Balloon heater
- Pyrrolidine	- Balloon
- Carboxymethoxy amine ½ HCl	- Refrigerant
- Methanol	- Rotavapeur
- Ethyl acetate	- Decanting funnel
- HCl	
- NaOH	
- $Na_2 SO_4$	
- Distilled water	

c- Experimental protocol

628.9 mg (2 mmol) progesterone (Sigma-Aldrich) and 284.5 mg (4 mmol) pyrrolidine are dissolved in 20 ml methanol. After 5 min with stirring, a yellowish precipitate (progesterone 3-(N-pyrrolidil) enamine) was formed. 218.6 mg (2 mmol) of carboxymethoxy amine ½ HCl is added to this stirred reaction mixture. The mixture is heated at 50-60°C for 5 min.

After cooling, the resulting solution (discolored and homogeneous) is evaporated under vacuum using the rotavapor, and the residue obtained is taken up in 40 ml of distilled water. This aqueous phase is then acidified to pH 2 with concentrated HCl solution, before being extracted 3 times with 50 ml ethyl acetate.

The recovered organic phase is, in turn, extracted 3 times with 50 ml of a 4% (w/v) NaOH solution. This alkaline solution is washed with 100 ml ethyl acetate, then cooled in an ice bath. Its pH is adjusted to 2 using concentrated HCl solution to precipitate the reaction product (P3-CMO). The resulting precipitate is extracted in ethyl acetate. The organic phase thus obtained is washed with distilled water, dried over $Na_2 SO_4$, then evaporated in vacuo using the rotavapor.

Recrystallization of the reaction product (P3-CMO) is possible after hot solubilization of the recovered residue in a methanol/distilled water mixture (v/v, 10 ml) and cooling (4°C).

d- Physicochemical analysis

The P3-CMO synthesis reaction was monitored by thin-layer chromatography using silica gel as the stationary phase and ethyl acetate/methanol (8/2; v/v) as the eluent.

The results obtained after UV revelation (Figure 15) show the disappearance of the starting reagent and the appearance of a new product with an R_f of 0.12 corresponding to that of
P3-CMO according to the literature (Mitsuma *et al.*, 1987).

Figure 15: *Silica gel TLC analysis of synthesized P3-CMO*
R being the initial reactant; P being the reaction product
The elution system used is a mixture of ethyl acetate and methanol (8/2; v/v).

The melting point of the synthesized product (P3-CMO) was evaluated at 169°C using the "*Electrothermal-9300 Digital Melting* Points *Apparatus*" supplied by UPR. This result is in perfect agreement with that reported in the literature, which stipulates a melting point for this product in the 168-170°C range (Allen and Redshaw, 1978).

As with P11α-HS, and in order to better elucidate the chemical structure of P3-CMO, a full analytical study was carried out at the CNRST UATRS. This study included the following analyses: IR spectrometry, NMR[1] H and mass spectrometry.

In particular, the IR spectrum recorded for the synthesized P3-CMO shows a broad band at 3429 cm^{-1} corresponding to the valence vibrations v_{O-H} of the carboxyl function, and an intense band at 1736 cm^{-1} corresponding to the valence vibrations $v_{C=O}$ of the carbonyl acid function (Figure 16).

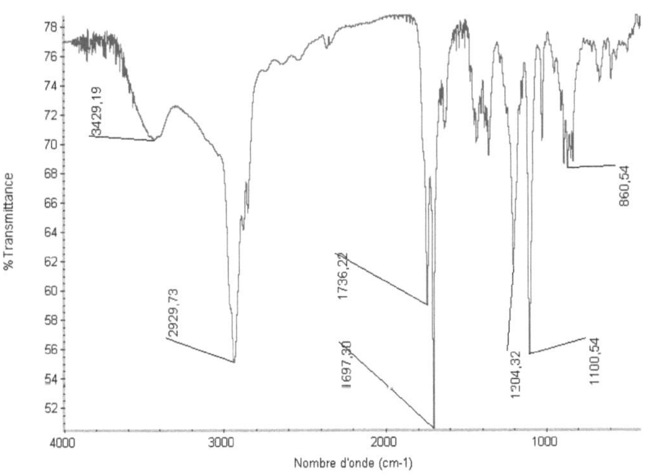

Figure 16: *IR spectrum of synthesized P3-CMO*
Solid sample; KBr

The[1] H NMR spectrum of the synthesized P3-CMO taken in deuterated chloroform shows, in addition to the signals due to the progesterone protons, a singlet at 4.69 ppm attributed to the protons of the (-CH$_2$ -) group linked to the oxime function, as well as a variation in the chemical shift corresponding to the protons of C$_4$ from 5.73 to 6.45 ppm, due to the insertion of a more electronegative chemical group at C$_3$ (Figure 17).

On the other hand, the mass spectrum of the synthesized P3-CMO in particular shows a molecular peak at m/z = 388.09 corresponding to [M + H] of this product (Figure 18).

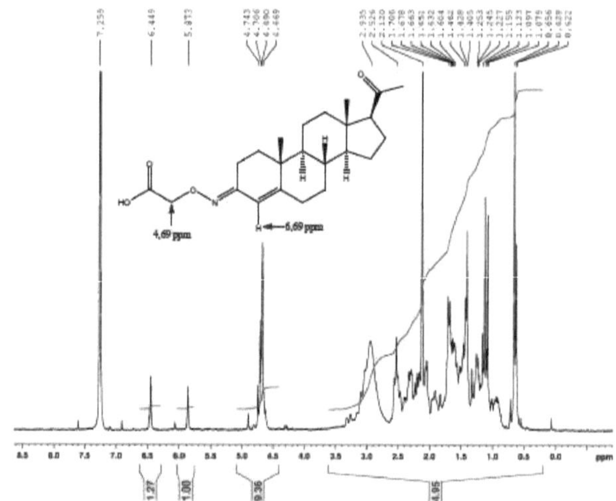

Figure 17: *^{1}H NMR spectrum of synthesized P3-CMO*
Sample taken in deuterated chloroform

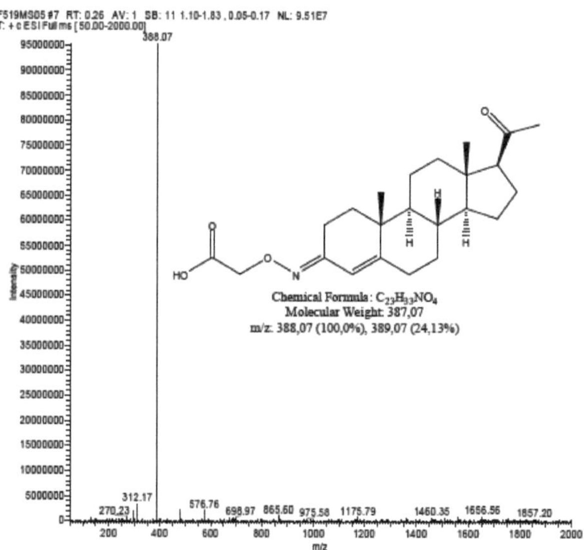

Figure 18: *Mass spectrum of synthesized P3-CMO*
Electrospray ionization in positive mode

All these results, obtained by means of IR, NMR[1] H and mass spectroscopy, confirm that the product synthesized is progesterone 3-(O-caroboxy-methyl) oxime (P3-CMO). The synthesis yield is estimated at 74%.

2. Coupling progesterone to BSA

After several trials involving the phenomenon of reagent solubility and solvent miscibility, we opted for the so-called "ester-activated" method to couple progesterone with BSA. Indeed, this method has been described in several research works as the method of choice for coupling steroids with carrier proteins (Bacigalupo *et al.*, 1987; Basu *et al.*, 2003; Basu *et al.*, 2006). It involves firstly preparing an ester-activated derivative of the steroid from the carboxyl-functional derivative and then, in a second step, coupling this ester-activated derivative with the carrier protein (BSA).

The choice of organic solvents and their proportions was based on their miscibility with water and their ability to solubilize the various reagents.

2-1. Preparation of ester-activated progesterone derivatives

Ester-activated progesterone derivatives were prepared, from the carboxyl-functional derivatives we synthesized, according to the method described by Basu *et al.*, (2006).

a- Reaction scheme

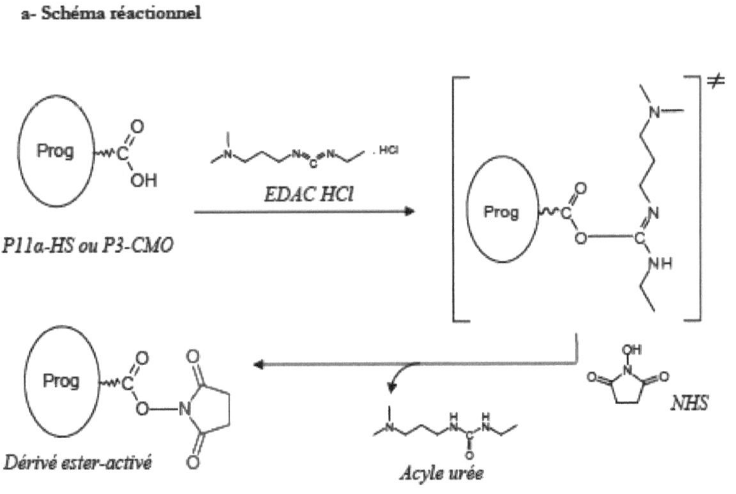

a- Schéma réactionnel

b- Materials required

Reagents	*Special equipment*
- Dimethylformamide (DMF) - 1,4-Dioxane - Distilled water - P11α-HS - P3-CMO - EDAC-HCl - *N-hydroxysuccinimide* (NHS)	- Two-way stirrer - Centrifuge

c- Experimental protocol

In a solution of 400 µl dimethylformamide (DMF) and 400 µl 1,4-dioxane, 12 mg (~ 0.031 mmol) of the synthesized carboxyl-functional progesterone derivative (P11α-HS or P3-CMO) is dissolved. To this is added 200 µl of a distilled water solution (freshly prepared) containing 12 mg of *N-hydroxysuccinimide* (NHS) and 24 mg of 1-ethyl-3-(3-dimethyl aminopropyl) carbodiimide hydrochloride (EDAC-HCl).

After *vortex* homogenization, the reaction mixture is left at room temperature with gentle stirring overnight.

Centrifugation at 15,000× g for 5 min then removes the urea acyl precipitate formed during the reaction, and clarifies the solution containing the ester-activated derivative prepared.

2-2. Coupling with BSA

Each of the two ester-activated progesterone derivatives, prepared as described above, was coupled to BSA to form an immunogen. The aim is to prepare two immunogens whose difference lies in the binding position of progesterone with BSA (3 or 11).

Thus, the ester-activated progesterone derivatives prepared were coupled with BSA according to the method described by Basu *et al.* (2006), which we have slightly adapted to our laboratory conditions. The experimental procedure followed is detailed as follows:

a- Reaction scheme

Ester-activated progesterone derivative

Bovine serum albumin

Progesterone-BSA" immunogen

b- Materials required

Reagents	Special equipment
- Ester-activated progesterone derivatives	- Magnetic stirrer
- Bovine serum albumin (BSA)	- Two-way stirrer
- Sodium borate (Na B O)$_{247}$	- pH meter
- Phosphate-buffered saline (PBS)	
- Dialysis bag (15,000 MW)	

c- Experimental protocol

The ester-activated progesterone derivative prepared in solution is gently added to 4 ml of a solution containing 42 mg (0.625 μmol) BSA in sodium borate buffer (0.1 M, pH 8) and kept at 4°C in an ice bath with continuous stirring.

The reaction mixture is then incubated overnight at 4°C. The resulting progesterone-BSA conjugate is dialyzed against 2× 5 l PBS buffer (pH 7.4) overnight at 4°C. It thus becomes the progesterone immunogen. It can be stored for several months at -20°C.

3. Characterization of prepared immunogens

In order to characterize the immunogens prepared and estimate the yield of coupling reactions, we used spectrophotometry and electrophoresis as analytical techniques. The results obtained are generally qualitative, although they can also be used to quantify the degree of coupling, as reported in several research studies (Yatsimirskaya *et al.*, 1993; Kothari *et al.*, 1995).

3-1. Spectrophotometric analysis

Using a UV-visible spectrophotometer (*Unicam UV-500),* we first determined the molar extinction coefficients of progesterone and BSA at 248 and 280 nm, their respective characteristic wavelengths (Figure 19).

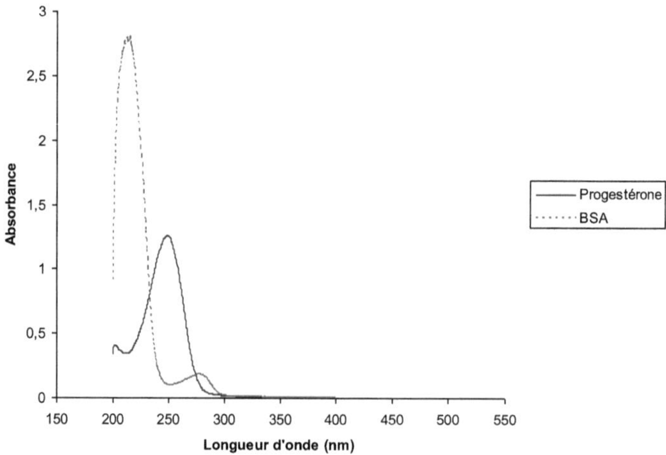

Figure 19: *UV-Visible spectra of progesterone and BSA in PBS*

For this purpose, we prepared a standard range of progesterone in PBS with concentrations ranging from 2 to 60 µg/ml, and a standard range of BSA under the same conditions but with concentrations ranging from 100 to 1000 µg/ml.

The results obtained, after reading the optical density of the various standards at 248 and 280 nm (Figure 20), enabled us to evaluate the molar extinction coefficients of progesterone at 1.44×10^4 and 8.74×10^2 M^{-1} cm^{-1} , and of BSA at 2.01×10^4 and 3.77×10^4 M^{-1} cm^{-1} , respectively at 248 and 280 nm (Table 2).

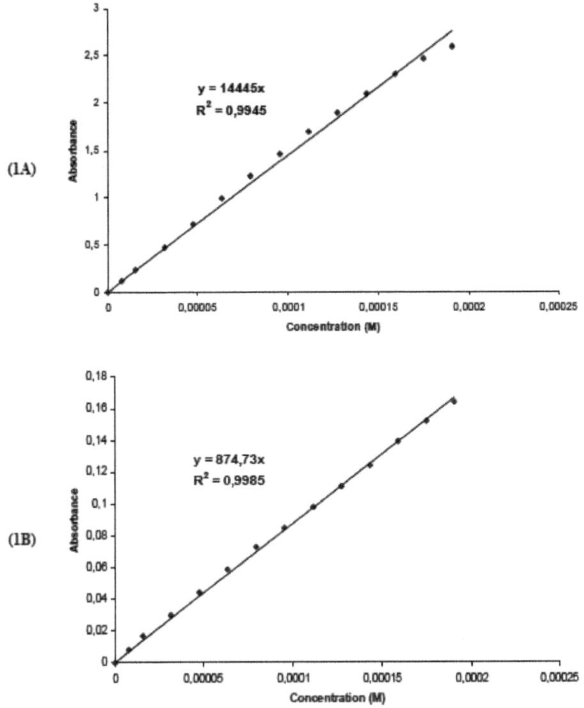

(1A)
Absorbance
y = 14445x
R² = 0,9945
Concentration (M)
(1B)
Absorbance
y = 874,73x
R² = 0,9985
Concentration (M)

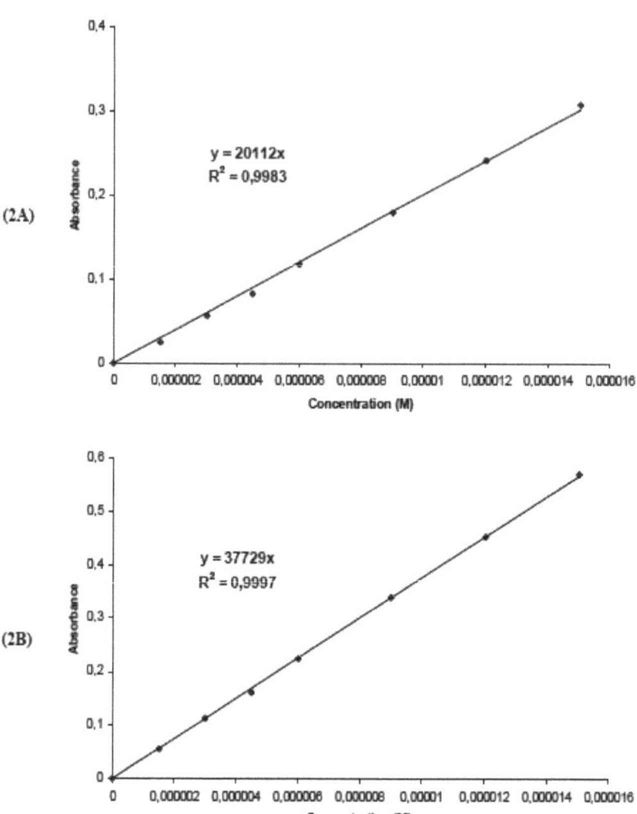

Figure 20: *Calibration ranges [OD = f (C)] for progesterone (1) and BSA (2) at 248 nm (A) and 280 nm (B)*

⇓

Table 2: *Molar extinction coefficients (M^{-1} cm^{-1}) for progesterone and BSA at 248 and 280 nm*

	248 nm	280 nm
Progesterone	1.44×10^4	8.74×10^2
BSA	2.01×10^4	3.77×10^4

Referring to these graphically determined molar extinction coefficients and measuring the absorbance at 248 and 280 nm of a solution containing approximately 250 µg/ml of the immunogen (conjugate) prepared in PBS, we were able to calculate the molar concentrations of progesterone (C_p) and BSA (C_B) for each immunogen using the following system of equations with two unknowns:

$$\begin{cases} A_T^{248} = A_P^{248} + A_B^{248} \\ A_T^{280} = A_P^{280} + A_B^{280} \end{cases} \Rightarrow \begin{cases} A_T^{248} = \varepsilon_P^{248} l C_P + \varepsilon_B^{248} l C_B \\ A_T^{280} = \varepsilon_P^{280} l C_P + \varepsilon_B^{280} l C_B \end{cases} \Rightarrow \begin{cases} A_T^{248} = \varepsilon_P^{248} C_P + \varepsilon_B^{248} C_B \\ A_T^{280} = \varepsilon_P^{280} C_P + \varepsilon_B^{280} C_B \end{cases}$$

with :

A_T^{248}	*Absorbance of conjugate at 248 nm*
A_T^{280}	*Absorbance of conjugate at 280 nm*
A_P^{248}	*Absorbance of progesterone at 248 nm*
A_P^{280}	*Absorbance of progesterone at 280 nm*
A_B^{248}	*Absorbance of BSA at 248 nm*
A_B^{280}	*Absorbance of BSA at 280 nm*
ε_P^{248}	*Molar extinction coefficient of progesterone at 248 nm*
ε_P^{280}	*Molar extinction coefficient of progesterone at 280 nm*
ε_B^{248}	*Molar extinction coefficient of BSA at 248 nm*
ε_B^{280}	*Molar extinction coefficient of BSA at 280 nm*
C_p	*Molar concentration of progesterone*
C_B	*Molar concentration of BSA*
l	*Optical path (1 cm)*

The solution of this system of equations with two variables (C_p and C_B) enabled us to estimate the coupling yield ($R=C/C_{pB}$) at 21 progesterone molecules per BSA molecule for the immunogen prepared from P11α-HS, and at 24 progesterone molecules per BSA molecule for the immunogen prepared from P3-CMO. The quantities prepared of these two immunogens are then respectively of the order of 46 mg for the former and 47 mg for the latter.

These results confirm the success of the various steps undertaken to prepare our immunogens, since the estimated coupling efficiencies exceed the critical value of 20 ($R > 20$) (Yatsimiriskaya *et al.*, 1993; Kothari *et al.*, 1995). Our immunogens are thus ready for use in the production of polyclonal anti-progesterone antibodies.

3-2. Electrophoretic *analysis*

With the same aim of characterizing the immunogens prepared and examining the smooth running of the coupling reactions, electrophoretic analysis of the conjugates resulting from these reactions was carried out under denaturing conditions on a 12% polyacrylamide gel. The results obtained are shown in Figure 21.

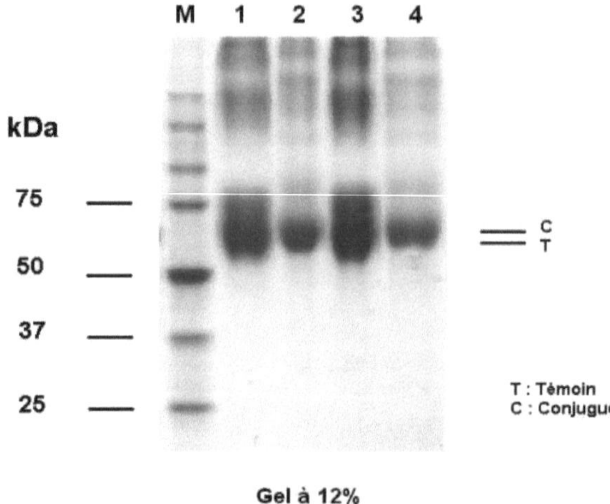

Figure 21: *Electrophoretic profiles of prepared immunogens on 12% polyacrylamide gel under denaturing conditions (PAGE/SDS)*
M: Molecular weight markers; 1 and 3: BSA; 2: Immunogen prepared from P11a-HS ;
4: Immunogen prepared from P3-CMO

These results confirm those obtained by spectrophotometry, insofar as they show that the molecular weight of the prepared immunogens (C) is significantly higher than that of the BSA control (T). This is a clear indication of the binding of progesterone molecules (MW ~ 400 g/mol) to BSA (MW = 66500 Da), resulting in an increase in the molecular weight of the conjugate.

II - IMMUNIZATION AND PREPARATION OF ANTIBODY EXTRACTS

1. Choice of animals

Most of the studies carried out on the development of progesterone immunoassay procedures point to the use of polyclonal antibodies produced in rabbits (Kothari and Pillai, 1998; Darwati *et al.*, 2006; Samuel *et al.*, 2006). Indeed, this animal has long been sought after for the production of polyclonal antibodies due to the many advantages it offers in this field, such as its ability to produce large quantities of antibodies thanks to its developed immune system, its rapid adaptation to laboratory rearing conditions, and its easy handling, particularly when injecting antigens or sampling blood.

With the same aim of developing progesterone immunoassay kits, other research has reported the use of mouse monoclonal antibodies (Christofidis *et al.*, 2006). These have a higher specificity than polyclonal antibodies. However, their field of application remains limited due to the technical difficulties encountered during their preparation.

On the other hand, to our knowledge, no study on the development of progesterone immunoassay kits has mentioned the use of polyclonal chicken antibodies. And yet, the fields of application of these antibodies are expanding day by day.

Chicken egg polyclonal antibodies (IgY) represent a very interesting alternative to rabbit polyclonal antibodies (IgG). Not only do they allow us to obtain 5 to 10 times more final material than rabbits, but they are especially interesting because of their phylogenetic distance from bovine, caprine or mammalian proteins in general. These antibodies therefore generate less background noise than polyclonal rabbit antibodies when used for mammalian applications, and in some cases enable better titers to be obtained than in a mammalian species.

For these reasons, we have opted to use both rabbits and hens to produce, against the same immunogens, two types of polyclonal anti-progesterone antibodies (IgG and IgY). These antibodies will be compared in terms of affinity and specificity, for use in progesterone immunoassay kits.

2. Animal immunization

Each of the two immunogens prepared was used to immunize two rabbits (1.5 kg) and two hens (1.5 kg). These animals were housed separately for acclimatization one week prior to immunization, during which time they were used to prepare pre-immune antibody extracts.

The immunization protocol followed is a conventional short-term protocol that has proven effective on several occasions. It can be summarized as follows (Figure 22):

Two milligrams of the "progesterone - BSA" immunogen prepared in 500 µl of PBS are mixed with 500 µl of Freund's incomplete adjuvant (Sigma-Aldrich). The mixture is stirred for 10 min (until a viscous white liquid is obtained), then injected subcutaneously in rabbits or intramuscularly in hens. The injections are made at different points on the animal's body.

After 21 days, a second booster injection with 2 mg of immunogen in 1 ml of physiological water (0.09% NaCl; w/v) is performed. Seven days later, 60 ml of rabbit blood or 6 hen eggs are collected.

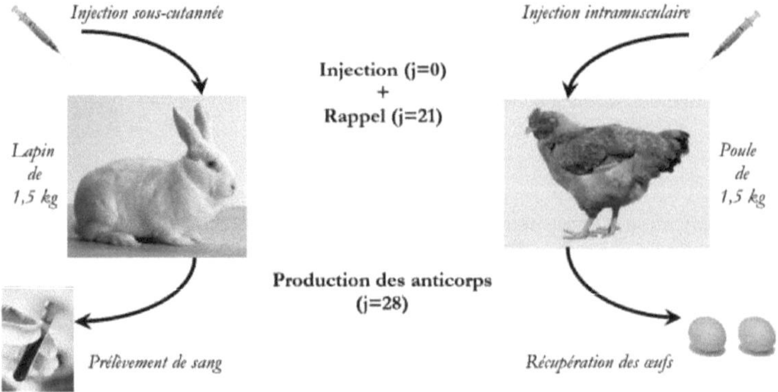

Figure 22: *Schematic representation of immunization protocols of rabbits and hens with prepared immunogens*

3. Preparation of antibody extracts

3-1. *Purification of total IgG*

Blood collected from rabbits is placed in a 37°C oven for 1 h, then refrigerated at 4°C overnight. After removal of the coagulum, the serum is recovered by centrifugation at 8000× g for 15 min. Sodium azide 0.02% (w/v) is added to the supernatant to form the antiserum.

Ammonium sulfate at 45% saturation is gently added to this antiserum. After 1 h precipitation in an ice bath (~ 4°C) with gentle agitation, centrifugation (15,000× g for 45 min) recovers a protein extract containing all polyclonal antibodies (total IgG). This extract is resuspended in a small volume of phosphate buffer (30 mM, pH 7.4), then dialyzed against the same buffer overnight at 4°C.

The dialyzed extract is then carefully introduced into a DEAE cellulose column (3× 20 cm) previously equilibrated with phosphate buffer (30 mM, pH 7.4). Under these pH and ionic strength conditions, IgG is positively charged. They are therefore recovered from the dead volume of the eluted column at a flow rate of 10 ml/h of phosphate buffer (30 mM, pH 7.4). This is referred to as negative chromatography. The recovered fractions (2 ml) are then analyzed by electrophoresis, protein assay and ELISA.

An example of the yields obtained after each purification step is summarized in Table 3. The values shown are the averages obtained for the purification of IgG from two rabbits immunized with the same immunogen.

Table 3: *Purification of total IgG from antisera*

	Volume (*ml*)	Concentration protein (*mg/ml*)	Protein totals (*mg*)	Activity (*U/ml*)	Specific activity (*U/mg*)	Total activity (*U*)	Purification factor (*times*)	Rdt (%)
EB	50	90,16	4508	14,26	0,158	712,26	1	100
AS	20	43,75	875	27,09	0,619	541,62	3,9	76
DEAE	26	9,24	240,24	17,03	1,843	442,76	11,6	62,1

EB: Gross extract
AS: Precipitation with ammonium sulfates (45%)
DEAE : Ion exchange chromatography (DEAE cellulose)

The activities shown in Table 2 are expressed in ELISA units per unit of volume (activity) or per quantity of protein (specific activity). Protein concentrations are determined by the Bradford method, with reference to the calibration curve.

3-2. *Purification of total IgY*

Polyclonal chicken antibodies (IgY) were recovered from egg yolks using the method described by Polson *et al.* (1980). In this method, the egg yolks (carefully separated from the whites) are placed in a test tube to which a solution containing phosphate buffer (30 mM, pH 7.4) and 0.01% (w/v) sodium azide is added in equal volume. The mixture is then mixed with 3.5% (w/v) polyethylene glycol 6000 and left to precipitate for 1 h under gentle agitation. After centrifugation (15,000× *g* for 15 min), the supernatant is precipitated a second time with the amount of polyethylene glycol 6000 required to reach a concentration of 12% (w/v).

The pellet containing total IgY recovered after centrifugation (15,000× *g* for 15 min) is resuspended in a small volume of phosphate buffer (0.030 M, pH 7.4) then dialyzed against this same buffer overnight at 4°C.

The dialyzed extract is then carefully introduced into a DEAE cellulose column (3× 20 cm) previously equilibrated with phosphate buffer (30 mM, pH 7.4). Under these pH and ionic strength conditions, the negatively-charged polyclonal chicken antibodies (IgY) cling to the column. The column is thoroughly washed with phosphate buffer (30 mM, pH 7.4) before being eluted by a linear gradient from 30 to 300 mM of phosphate buffer (pH 7.4) at a flow rate of 10 ml/h. The recovered fractions (2 ml) are then analyzed by electrophoresis, protein assay and ELISA.

Table 3 shows an example of the yields obtained for the different purification steps of total IgY. The values shown are the averages obtained for the purification of total IgY from two hens immunized with the same immunogen.

The activities shown in Table 4 are expressed in ELISA units per unit of volume (activity) or per quantity of protein (specific activity).

Table 4: *Purification of total IgY from egg yolks*

	Volume (*ml*)	Concentration protein (*mg/ml*)	Protein totals (*mg*)	Activity (*U/ml*)	Specific activity (*U/mg*)	Total activity (*U*)	Purification factor (*times*)	Rdt (%)
EB	120	-	-	-	-	-	-	-
PEG	20	160,24	3204,75	65,21	0,407	1304,33	1	100
DEAE	32	15,04	481,28	29,43	1,956	941,38	4,8	72,1

EB: Gross extract
PEG: fractional precipitation with PEG 6000 (3.5 - 12%)
DEAE : Ion exchange chromatography (DEAE cellulose)

3-3. Biochemical analysis by Dot-Blot

This analysis was carried out qualitatively to highlight the presence of anti-progesterone antibodies and assess their affinity for progesterone. It involved both total IgG and total IgY extracts.

Assays were performed on nitrocellulose membranes, using progesterone coupled with BSA (immunogenic) and progesterone coupled with KLH as antigens. It should be noted that the coupling of progesterone with KLH was carried out using the same protocol as described for the coupling of this hormone with BSA (ester-activated method). The antigen resulting from this "progesterone - KLH" coupling is used to assess the presence of progesterone-specific antibodies.

The results obtained, shown in Table 5, clearly demonstrate the production of polyclonal anti-progesterone antibodies in all immunized animals (rabbit and hen).

Table 5: *Dot-Blot analysis of prepared antibody extracts*

	Pre-immun	Immune extract
Rabbit (IgG)		
A		A
B		B

Hen (IgY)

Series A: the antigen used is "progesterone-BSA".
Series B: the antigen used is "progesterone-KLH".

III - PURIFICATION OF ANTI-PROGESTERONE ANTIBODIES

Anti-progesterone antibodies were purified from antibody extracts (total IgG and total IgY) by affinity chromatography using Affigel 10 agarose gel (Biorad) as support and progesterone as ligand.

1. Chromatography support activation

Progesterone activation of the chromatography support was carried out according to the following protocol: 5 ml agarose gel (Affigel 10) in 5 ml isopropanol was incubated with 60 µl ethylene diamine in 300 µl DMF for 30 min at room temperature with gentle agitation. This step consists in substituting the NHS groups of the Affigel with a side chain bearing a primary amine group (-NH$_2$). After several washes, the gel is taken up to equal volume in isopropanol. The ligand (a carboxyl-functionalized progesterone derivative) is then coupled to the support using the same protocol previously described for coupling progesterone to BSA (ester-activated method).

2. Purification of progesterone-specific antibodies

Extracts containing the polyclonal antibodies to be purified (total IgG or total IgY) are incubated with the *batch-activated* gel overnight at 4°C with gentle agitation. After washing with PBS the gel introduced into a column (1 × 10 cm), elution is carried out with a 0.1 M glycine solution (pH 2.5). Fractions (1 ml) containing anti-progesterone antibodies are collected in tubes containing 100 µl of Tris-HCl solution (1 M, pH 9) (Figure 23).

The column can be reused around 10 times after regeneration with a glycine solution (0.1 M, pH 2.5) followed by a PBS wash. The column is stored in the presence of an anti-microbial agent (0.02% sodium azide).

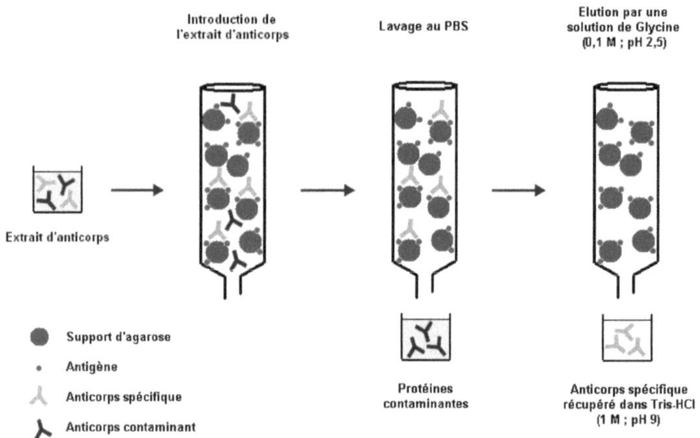

Figure 23. *Schematic representation of the various steps involved in the purification of polyclonal anti-progesterone antibodies by affinity chromatography.*

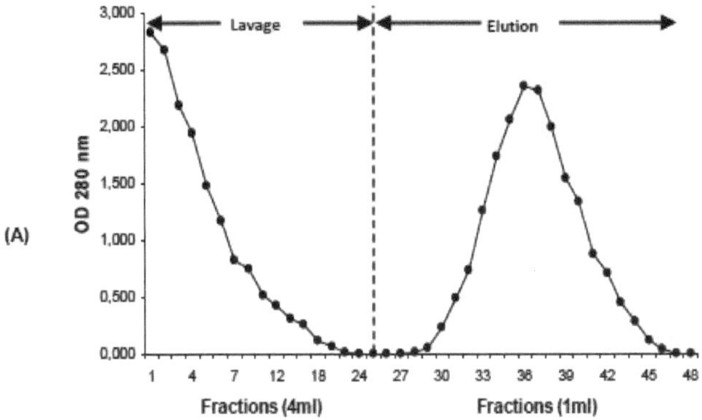

Purification d'anticorps par chromatographie d'affinité

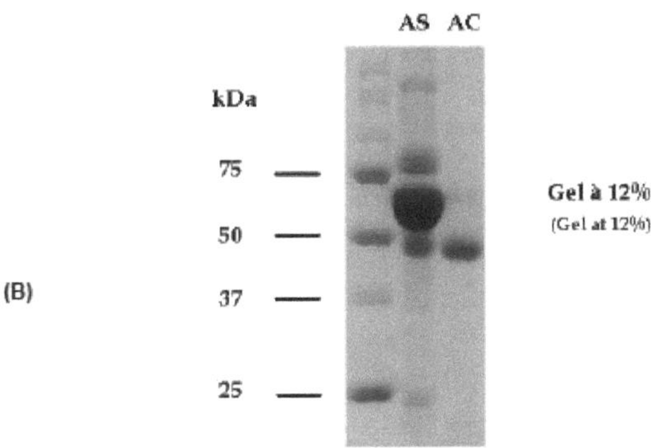

AS : Antisérum brut/ Crude antiserum (0,3 µl)
AC : Anticorps purifié/ Purified antibody (15µg)

Figure 24: *Elution profile of rabbit anti-progesterone antibodies (A) and electrophoretic profile showing purity of purified antibodies (B)*

Figure 24 shows the elution profile of rabbit anti-progesterone (IgG) antibodies purified on an affinity chromatography column. The purity of fractions recovered after elution is verified by electrophoresis.

Purification yields are shown in Table 6. The values shown represent the averages obtained during purification of antibody extracts (total IgG) prepared from the antisera of two rabbits immunized with the same immunogen.

Table 6: *Purification of anti-progesterone antibodies from extracts of total IgG*

51

	Volume (*ml*)	Concentration protein (*mg/ml*)	Protein totals (*mg*)	Activity (*U/ml*)	Specific activity (*U/mg*)	Total activity (*U*)	Purification factor (*times*)	Rend. (*%*)
EG	26	9,24	240,24	17,03	1,843	442,76	1	100
CA	12	1,12	13,44	28,92	25,820	347,02	14	78,3

EG: Total IgG extract
CA: Affinity chromatography

As with IgG extracts, IgY extracts were purified to recover progesterone-specific antibodies produced in hens. For this, we used the same protocol described above for the purification of anti-progesterone IgG.

Figure 25 shows the elution profile of purified chicken anti-progesterone (IgY) antibodies on the affinity chromatography column. Verification of the purity of the fractions recovered after elution is carried out as previously mentioned by electrophoresis.

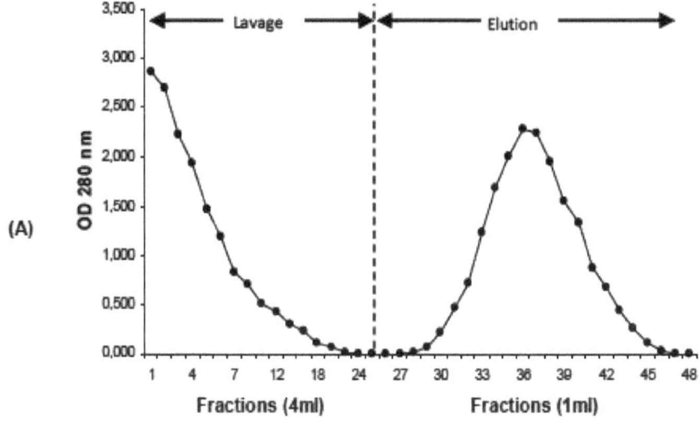

Purification d'anticorps par chromatographie d'affinité

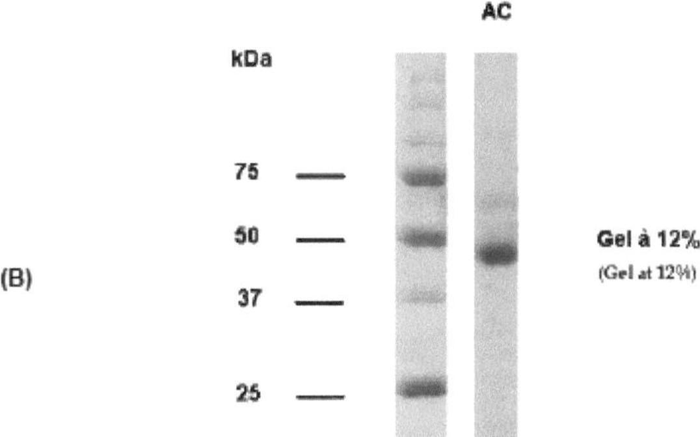

AC : Anticorps purifié/ Purified antibody (15µg)

Figure 25: *Elution profile of chicken anti-progesterone antibodies (A) and electrophoretic profile showing purity of purified antibodies (B)*

Purification yields are shown in Table 7. The values shown represent the averages obtained when purifying antibody extracts (total IgY) from two hens immunized with the same immunogen.

Table 7: *Purification of anti-progesterone antibodies from extracts of total IgY*

Volume (*ml*)	Concentration protein (*mg/ml*)	Protein totals (*mg*)	Activity (*U/ml*)	Specific activity (*U/mg*)	Total activity (*U*)	Purification factor (*times*)	Rend. (%)

53

EY	25	10,57	264,25	18,95	1,792	473,53	1	100
CA	12	1,46	17,52	30,01	17,680	309,75	9,8	65,4

EY: Total IgY extract
CA: Affinity chromatography

The purified polyclonal anti-progesterone antibodies (IgG and IgY) are supplemented with 0.02% (w/v) sodium azide, then stored in aliquots at -20°C until use. They will subsequently be used in the development of progesterone immunoassay kits.

On the other hand, in order to assess the affinity and specificity of these locally produced polyclonal anti-progesterone antibodies, trials are currently underway to study cross-reactions between these antibodies and a number of compounds presenting structural analogies with progesterone (testosterone, cortisol, estradiol, pregnenolone, deoxycorticosterone, 5 β-dihydroprogesterone, 6 β-dihydroprogesterone, 20 α-dihydro-progesterone, 16 α-hydroxyprogesterone, 17 α-hydroxyprogesterone).

We also expect this study to demonstrate the value of using polyclonal chicken antibodies (IgY) to increase the specificity and sensitivity of immunoassay systems intended for use in mammals, as is the case with the assay system we are currently developing.

PART 2: PREPARATION AND VALIDATION OF THE LOCAL PROGESTERONE STANDARD RANGE

The standard range is an essential component of an immunoassay kit. It consists of standard solutions with a known concentration of antigen (analyte). These standards are used to plot a calibration curve, from which extrapolation can be performed to determine the unknown concentration of a sample.

Standards are generally prepared in a matrix devoid of the substance to be assayed (antigen). This matrix must reproduce exactly the conditions in which the antigen to be assayed is found. The concentration range of standards may vary from one standard range to another. However, this range must cover physiological and/or pathological values (Lupi-Chen *et al.*, 1999).

In this part of our work, we describe the preparation of a progesterone standard range to be used for the specific determination of this hormone in goats. This standard range, prepared locally, was methodologically validated against a commercial standard range.

I - MATRIX PREPARATION

1. Matrix selection

The matrices used to prepare progesterone standards differ from one manufacturer to another. However, the majority of progesterone immunoassay kits on the market use human physiological fluid (serum or plasma) as the matrix.

These kits are perfectly suited to the immunoassay of progesterone in humans, but less so in animals. Indeed, although there may be similarities between human and animal physiological fluids, they remain different environments, which may explain the background noise and interference frequently encountered when using a human progesterone immunoassay kit to measure this hormone in animals.

Consequently, the choice of matrix must be dictated by the reaction environment of the antigen to be assayed (pH, ionic strength, protein composition, etc.), and therefore by the nature of the desired application of the assay system to be developed.

All these reasons prompted us to use goat serum as the assay matrix for our standard range, given that the immunoassay system we plan to produce is dedicated to the assay of progesterone in goats. The billy goat was chosen because its serum contains much less progesterone than that of the goat, and therefore requires less rigorous processing.

2. Matrix processing

2-1. Serum depletion

This technique removes all traces of steroids from serum by physical adsorption on activated carbon. It also eliminates certain proteins that can complex steroids. This is an essential step in the preparation of the matrix for the assay of progesterone or similar steroid hormones.

2-2. Materials required

Reagents	Special equipment
- Powdered activated carbon	- Balance
- Filtration membrane (0.6 - 0.1 μm)	- Two-way stirrer
- Bradford reagent	- Filtration system (Millipore)
- Bovine serum albumin (BSA)	- Centrifuge
- RIA progesterone assay kit	- Gamma counter

2-3. Experimental protocol

Goat serum (courtesy of our partner INRA Tangiers) is depleted with activated charcoal at 5 g / 100 ml (w/v) for 48 h at 4°C. After centrifugation at 9,000× g for 30 min, the supernatant is successively filtered through 0.6, 0.4, 0.2 and 0.1 μm membranes to remove all fine charcoal particles.

The protein concentration of depleted serum is determined by the Bradford method, using BSA as the standard protein. The absence of any trace of progesterone is verified by direct radioimmunoassay of the extract obtained (depleted serum).

The results obtained (Table 8) testify to the effectiveness of the depletion protocol used, since it enabled us to prepare a goat serum devoid of progesterone and any other steroid that might interfere with the assay. This serum will be the matrix we use to prepare our standard range.

Table 8: *Characteristics of goat serum before and after depletion*

	Protein concentration (mg/ ml)	Progesterone concentration (ng/ ml)
Raw serum	94,51	0,182
Depleted serum	72,38	0,000

II - PREPARING THE PROGESTERONE STANDARD RANGE

1. Preparation of standards

Standards were prepared, in depleted goat serum (matrix) containing 10 mM sodium azide (NaN$_3$), from a stock solution of progesterone in ethanol (30 mg in 100 ml) whose concentration was verified spectrophotometrically at 248 nm. The molar extinction coefficient used is that of progesterone ($\varepsilon_p^{248nm} = 14400$ M^{-1} cm)$^{-1}$

From this stock solution, three 1:10 cascade dilutions were prepared using mixtures (distilled water/ethanol) with volume ratios of (1:1), (9:1) and (1:0) respectively. The concentrations of these three diluted solutions were in turn checked spectrophotometrically at 248 nm. These solutions were the starting points for the preparation of the various standards in our standard range, whose concentrations are shown in Table 9.

Table 9: *Target concentrations for local standards progesterone*

Standards	S_0	S_1	S_2	S_3	S_4	S_5	S_6
[Progesterone] (ng/ml)	0,000	0,390	0,800	1,640	5,300	15,90	30,00

2. Progesterone determination and calibration of standards

Progesterone concentrations in locally prepared standards were then assessed using Diasource's "PROG-RIA-CT KIP1458" radioimmunoassay kit (Figure 26), following the manufacturer's instructions.

In this way, progesterone contained in the standards (from the local standard range or the commercial standard range) competes with iodine-125-labeled progesterone against a given, limited number of anti-progesterone antibodies.

At the end of the incubation period, the amount of labelled progesterone bound to the antibody is inversely proportional to the amount of unlabelled progesterone present in the sample. The proposed methodology for separating free and bound fractions uses antibody-coated tubes.

The results obtained, after measuring the radioactivity linked to the coated tubes using a gamma scintillator, were used to check and calibrate the various standards in our standard range, before proceeding with its qualification and validation.

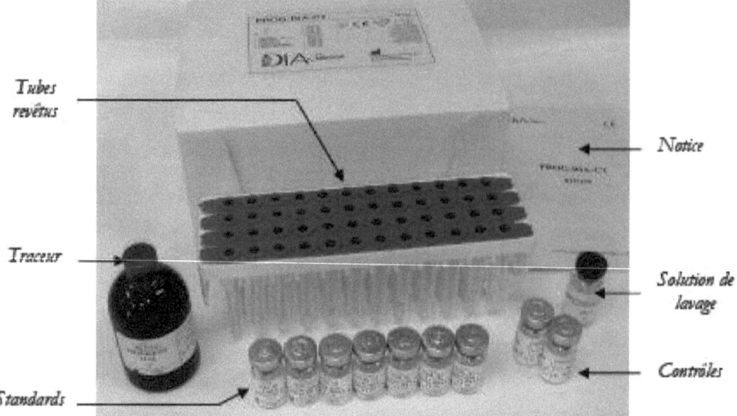

Figure 26: *Components of the progesterone radioimmunoassay kit*
(PROG-RIA-CT KIP1458, Diasourse)

III - VALIDATION OF THE PROGESTERONE STANDARD RANGE

Methodological validation of the standard range, prepared locally, was carried out against the standard range marketed with the Diasource progesterone radioimmunoassay kit (PROG-RIA-CT KIP1458).

1. Definition of validation

Validation is the process of demonstrating that any process or procedure used to dose a product actually leads to the expected result. It is defined as: the confirmation, by examination and the provision of objective evidence, that the particular requirements for a given intended use have been met (Caporal-Gautier *et al.*, 1992).

2. Validation criteria

Validation of a standard range is carried out in accordance with international standards (Caporal-Gautier *et al.*, 1992; Albrecht *et al.*, 2004), using several validation criteria which are either functional or statistical:

- *Functional criteria*: sensitivity, selectivity and specificity
- *Statistical criteria*: linearity, fidelity, accuracy and detection limit

These parameters define the ability of the standard range to give accurate and consistent results under specific operating conditions. The usual validation criteria are (Dutruc-Rosset, 1999) :

- Specificity - selectivity
- Fidelity (repeatability and reproducibility)
- The confidence interval
- Accuracy
- Linearity
- The response function

2-1. *Loyalty*

According to EEC Explanatory Note III/844/87, drawn up in 1989, precision expresses the closeness of agreement (degree of dispersion) between a series of measurements taken from multiple samples of the same homogeneous material under prescribed conditions. Fidelity provides an indication of errors due to chance. It comprises two parameters:

a- Repeatability (precision)

It expresses precision under identical conditions, known as repeatability conditions: same analyst, same equipment, same reagents, within a short time interval. Repeatability or precision is a dispersion parameter that expresses the minimum variability of an analytical procedure (Fabre, 1999, Albrecht *et al.*, 2004). Repeatability is generally expressed by the relative standard deviation RSD of repeatability (or intra-assay coefficient of variation):

$$CV = RSD = \frac{S}{x} \times 100$$

✓ *Reduced standard deviation (S) or standard deviation (SD):* $S = \sqrt{\dfrac{\sum (xi - \bar{x})^2}{n-1}}$

✓ *Within-test variance:* $S^2 = \dfrac{\sum (xi - \bar{x})^2}{n-1}$

with :

x_i : *individual value*

$\bar{x}$ *average of individual values*

n: number of measurements or tests

b- Reproducibility (intermediate precision)

It expresses precision under different conditions: analysts, equipment, laboratories, reagents, days (individual values are collected over at least 3 days). Reproducibility is a dispersion parameter that expresses the maximum variability of an analytical procedure (Caporal-Gautier *et al.*, 1992). Reproducibility is expressed in a similar way to repeatability, i.e. by the relative standard deviation RSD of reproducibility with respect to the mean $\bar{x}$ of individual values.

c- Minimum requirements

- *Repeatability: n = 6; n being the different measurements taken on each point.*
- *Reproducibility: k = 3; k being the different test groups.*

d- Experimental design

At least *k* groups of *n* trials are carried out; experimentally, the assay is carried out by 3 different manipulators (*k = 3*) on 3 different days carrying out 6 measurements (*n = 6*) (Table 10) :

Table 10: *Experimental conditions for methodological validation*

	Test groups		
	Day 1	Day 2	Day 3
	Group 1 = Operator 1	Group 2 = operator 2	Group 3 = Operator 3
Tests	Y_{11}	Y_{12}	Y_{13}
	Y_{21}	Y_{22}	Y_{23}
	...	...	...
	Y_{61}	Y_{62}	Y_{63}

e- Determining and calculating loyalty criteria

The different variances (intra-assay S_j^2 , repeatability S_r^2 , reproducibility SR^2 , inter-group S_g^2) are used to calculate the intra- and inter-assay coefficients of variation, which are used to estimate repeatability and reproducibility respectively:

** Intra-assay variance S_j^2 :* or dispersion within each measurement group :

$$S_j^2 = \frac{\sum_{i=1}^{n_j}(Y_{ij}-m_j)^2}{n_j-1}$$

with :

S_j : standard deviation within group j
i   : individual value index in group j
j   group index (1, 2, 3)

60

Y_{ij} *gross* dependent *value*
m_j *average of group j*
n_j : *number of measurements within the group*
n_j -1 : *number of degrees of freedom (ddl)*

** Repeatability variance S_r^2 :*

$$S_r^2 = \frac{\sum_{j=1}^{k}\left[(n_j-1)\times S_j^2\right]}{\left(\sum_{j=1}^{k} n_j\right)-k} \; ; \; k = 3 \; ; \; n_j = 6$$

If all groups perform the same number of measurements (n = 6), the repeatability variance

is: $S_r^2 = \dfrac{\sum_{j=1}^{k} S_j^2}{k}$

** Inter-group variance S_g^2 :*

It is given by the following formula, in the case where all groups perform the same

number of measurements (n = 6): $S_g^2 = \dfrac{\sum_{j=1}^{k}(m_j-\overline{m})^2}{k-1} - \dfrac{S_r^2}{n}$

with :

    S_r^2 : *repeatability variance*
    m_j : *average for group j*
    $\overline{m}$ *average of k group averages*

** Reproducibility variance S_R^2 :*

It includes both repeatability variance and inter-group variance. It is given by the formula:
$S_R^2 = S_r^2 + S_g^2$

** Coefficient of variation of repeatability (intra-assay) CV_r :* $CV_r = \dfrac{S_r}{m} \times 100$

** Coefficient of variation of reproducibility (inter-assay) CV_R :* $CV_R = \dfrac{S_R}{m} \times 100$

The acceptance limits for coefficients of variation (CV_r and CV_R) are of the order of 10% (Jardy and Vial, 1999).

2-2. Accuracy

Accuracy expresses the closeness of agreement between the experimental value and the accepted reference value. It combines both accuracy and precision. Accuracy provides an indication of systematic errors. The result can be expressed as a percentage of recovery in relation to known quantities. Percentage recovery is given by the following formula:

$$R_m\,(\%) = \frac{C_{exp}}{C_{th}} \times 100$$

with :

> C_{exp} *: experimental solution concentration*
> C_{th}*theoretical solution concentration*

It must be between 80 and 120% of the theoretical value to be determined, i.e. the concentration of each standard in the standard range.

2-3. Confidence interval

Alongside the standard deviation and coefficient of variation, the confidence interval must be determined for each concentration.

The confidence interval is given by: $\left[\overline{m} - 2\,\frac{SD}{\sqrt{n}}, \overline{m} + 2\,\frac{SD}{\sqrt{n}} \right] ; t = 2$

with :

> $\overline{m}$ *: medium*
> n *number of measurements*
> *SD: standard deviation*

2-4. Linearity

The linearity of a standard range is its ability, within a certain interval, to provide results directly proportional to the concentration of the substance to be determined in the sample. This is expressed by the correlation coefficient. Thus, by plotting the measured concentrations of the local standard against those of the commercial standard, we can establish a statistical relationship between the concentrations of the two standards. This correlation may be absent, linear or non-linear, and is expressed by the correlation coefficient R, which is given by the formula and can be deduced from the regression line :

$$R = \frac{\dfrac{1}{n}\sum_{i=1}^{n}(x_i - \overline{x})(y_i - \overline{y})}{\sqrt{\sum(x_i - \overline{x})^2 \, \sum(y_i - \overline{y})^2}} \; ; n = 6$$

with :

> x_i *: index of the individual value of the theoretical progesterone concentration of the standard range developed locally*

y_i : *index of the individual value of the experimental progesterone concentration of the standard range developed locally*

-1 ≤ R ≤ 1 in the general case
R = ±1 for an exact linear relationship
R = 0 in the absence of a linear relationship

2-5. Response function

The response function expresses, within the assay interval, the relationship between the response and the concentration of the substance in the sample.

3. Results of the validation of the local progesterone standard range

Three manipulators representing 3 different groups (1, 2 and 3) performed 6 determinations of each standard. Thus, the minimum conditions required for validation of our local standard range are met (k=3; n=6).

For all assays carried out, the commercial standard gave us a total percentage of tracer bound in the absence of progesterone (B0/T) of between 40 and 50% (a *sine qua non for* assay validity). The standard curve obtained is correct (Figure 27) and the internal controls are included in the experimental ranges defined by the manufacturer.

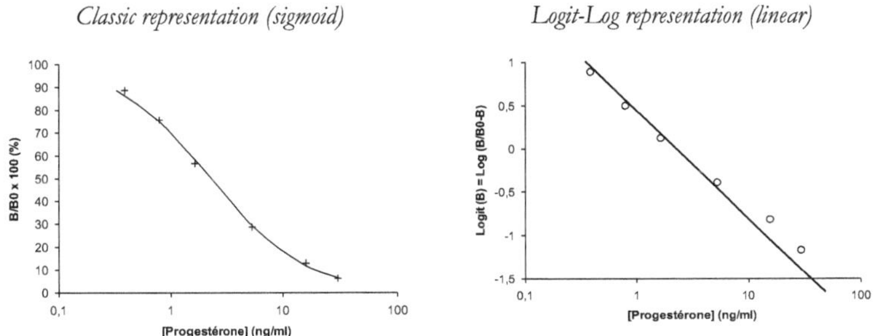

Figure 27: *Calibration curve for commercial progesterone assay kit*
PROG-RIA-CT KIP1458 from Diasource

This calibration curve, representing the percentage of tracer binding as a function of progesterone concentration (Figure 27), enabled us to estimate the concentration of this hormone in the various standards of our local standard range. The raw results obtained are shown in Table 11.

	S$_0$	S$_1$	S$_2$	S$_3$	S$_4$	S$_5$	S$_6$
Group 1	0,000	0,414	0,790	1,733	6,092	15,77	31,18
	0,000	0,330	0,925	1,886	5,843	15,74	30,78
	0,000	0,324	0,899	1,742	5,991	15,91	31,07
	0,000	0,408	1,017	1,743	5,932	15,45	30,30
	0,000	0,384	0,945	1,801	6,016	15,13	31,15
	0,000	0,351	0,827	1,820	5,765	15,37	29,81
Group 2	0,000	0,372	0,761	1,688	6,222	14,91	29,88
	0,000	0,402	0,743	1,778	5,287	15,54	29,45
	0,000	0,342	0,848	1,630	6,126	14,69	30,03
	0,000	0,351	0,877	1,715	5,720	14,08	30,11
	0,000	0,354	0,907	1,697	6,308	15,30	27,86
	0,000	0,417	0,795	1,753	5,851	15,30	29,30
Group 3	0,000	0,357	0,908	1,769	5,118	15,00	25,71
	0,000	0,348	0,890	2,053	5,208	16,01	31,12
	0,000	0,363	0,907	2,186	5,213	14,70	29,57
	0,000	0,405	0,903	1,891	5,093	15,16	29,74
	0,000	0,399	0,951	1,952	5,597	14,20	30,36
	0,000	0,423	0,853	2,091	4,994	14,38	26,59

Values represent concentrations in ng/ml.

The actual progesterone concentrations of the standards in our standard range were thus calculated. The mean values obtained are shown with the expected concentrations in Table 12.

On the basis of these results, the various fidelity criteria for the local progesterone standard range were determined. These criteria are shown in Table 13.

	Average concentration (ng/ml)			Concentration obtained (ng/ml)	Concentration expected (ng/ml)
	Group 1	Group 2	Group 3		
S_0	0,000	0,000	0,000	0,000	0,000
S_1	0,369	0,373	0,383	0,374	0,390
S_2	0,900	0,821	0,902	0,874	0,800
S_3	1,787	1,710	1,990	1,820	1,640
S_4	5,939	5,919	5,203	5,687	5,300
S_5	15,56	14,97	14,90	15,14	15,90
S_6	30,71	29,43	28,84	29,66	30,00

Table 13: *Fidelity criteria for the local standard range progesterone*

	S_r^2	S_g^2	S_R^2	CV_r (%)	CV_R (%)	R_m (%)
S_0	-	-	-	-	-	-
S_1	0,001124	-0,000136	0,000987	8,949	8,389	96,06
S_2	0,004026	0,001431	0,005458	7,253	8,445	109,34
S_3	0,009612	0,019333	0,028946	5,359	9,300	111,54
S_4	0,067971	0,164270	0,232241	4,584	8,473	107,31
S_5	0,263904	0,086135	0,350039	3,392	3,906	95,26
S_6	1,916287	0,591022	2,507310	4,666	5,337	98,89

with

S_r^2	*repeatability variance*
S_g^2	*inter-group variance*
S_R^2	*reproducibility variance*
CV_r	*intra-assay coefficient of variation*
CV_R	*inter-assay coefficient of variation*
R_m	*average recovery rate expressing accuracy*

We found that :

✓ The coefficients of variation for repeatability and reproducibility (CV_r and CV_R), calculated for the different standards in our local standard range, do not exceed the acceptance limit of around 10%. This testifies to the homogeneity of the variances between the 3 measurement groups and also within each group.

✓ The value of the inter-group variance of the standard S_1 is negative. This occurs when the inter-group variance (S_g^2) is not significant compared to the repeatability variance. This value can therefore be reset to zero.

✓ The recovery rates (R_m) expressing accuracy are between 80 and 120% for all the standards in our standard range. Consequently, differences between actual and expected progesterone concentrations in the standards are not significant.

The local progesterone standard range is thus statistically validated. Table 14 shows the mean concentrations and confidence intervals of the various standards in this validated standard range.

Table 14: *Concentrations and confidence intervals of the various standards in the local progesterone standard range*

	$\overline{m}$ (ng/ml)	SD	$\left[\overline{m}-2\dfrac{SD}{\sqrt{n}}, \overline{m}+2\dfrac{SD}{\sqrt{n}}\right]$
S_0	0,000	0,000	[0,000 - 0,000]
S_1	0,374	0,032	[0,359 - 0,389]
S_2	0,874	0,070	[0,841 - 0,908]
S_3	1,829	0,152	[1,757 - 1,901]
S_4	5,687	0,428	[5,485 - 5,889]
S_5	15,14	0,56	[14,87 - 15,41]
S_6	29, 66	1,52	[28,94 - 30,38]

with :

$\overline{m}$ *: medium*
n: number of measurements
SD: standard deviation

For greater accuracy, the linearity of our local progesterone standard range was verified by assessing the correlation between this range and that of the reference system with which it was validated. The results obtained are shown in Figure 28.

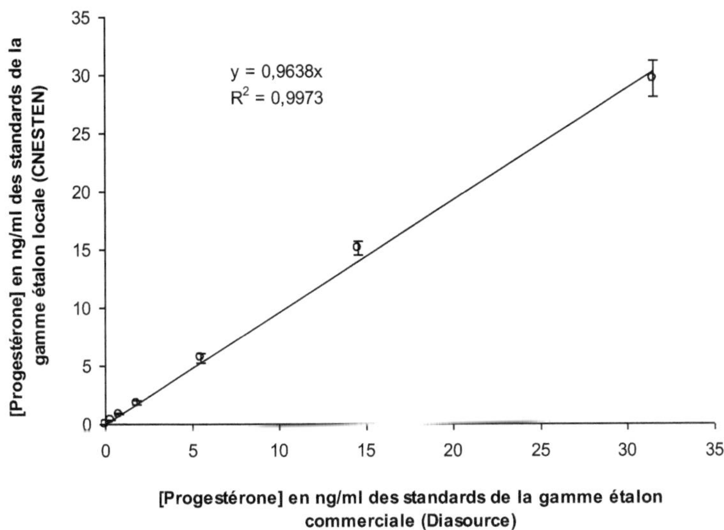

Figure 28: *Correlation between the local standard range and the commercial standard range (Diasource) for progesterone*

These results show a linear correlation between the concentrations of the local progesterone standard and those of the commercial standard (Diasource). This linearity is confirmed by the correlation curve, which shows a straight line whose equation is: "$y = 0.964\,x$". The value of the R^2 is evaluated at 0.99, testifying to perfect linearity.

After validation of our local progesterone standard range, the various standards making it up are stored, in aliquots, at -20°C until use. Trials are currently underway to study the stability of this standard range under different storage and handling conditions.

PART 3: PREPARATION OF RADIOACTIVE PROGESTERONE TRACERS

Most immunoassays use markers to reveal the immune complex formed by antigen and antibody. These markers are entities chemically linked (covalently or non-covalently) to an antigen (or antibody) molecule, and which deliver a quantitatively measurable signal. The tracer resulting from this association must behave in the same way as the ligand (antigen or antibody); in other words, it must retain the same physico-chemical and immunological properties after labeling.

Numerous markers are used in immunoassays (Pelizzola *et al.*, 1995). These are usually radioactive isotopes (in the case of radioimmunoassays) or enzymes (in the case of enzyme immunoassays). However, radioisotopes are still the most frequently used markers, due to their high sensitivity, predisposition to automation and, above all, ease of use (Basu *et al.*, 2003).

In a radioimmunoassay, the concentration of labelled antigen (tracer) is determined using a gamma counter (RIA-STAR or other) capable of measuring the signal emitted by the radioisotope used for labelling (iodine-125 or other). In addition, the choice of radioisotope and the way it is coupled to and labelled with the antigen will depend on the intended applications of the assay system to be developed.

In this part of our work, we'll be focusing on the preparation of radioactive progesterone tracers. These tracers, prepared by coupling progesterone with a molecule capable of being radiolabelled and labelling the resulting conjugates with iodine-125, will form part of the radioimmunoassay kit we plan to produce for measuring this hormone in goats.

I - CHOICE OF MARKING RADIOISOTOPE

An immunoassay marker is an entity chemically linked to an antigen or antibody molecule and delivering a quantitatively measurable signal.

Both radioactive isotopes are used in the manufacture of radiolabeled ligands:

- Iodine 125 (125I)
- Tritium (3H).

1. Advantages of iodinated ligands

- Iodine 125 (Na125I) is an inexpensive molecule with relatively low ionizing power, but with sufficient energy to emit measurable radiation.

- The methods for incorporating radioactive iodine involve simple chemistry, accessible to biochemists.

- The half-life of 125I (60 days) is 73 times shorter than that of 3H (12.3 years). Every 60 days, the specific radioactivity of iodinated waste is halved. Very quickly, this waste will no longer be radioactive.

- The electrochemical conversion efficiency of 125I on photo film is 3-10% (depending on the type of film), while that of tritium is less than 1%. The image will be obtained 200-500 times faster with an iodinated ligand than with 3H.

2. Incorporation of the 125I radioelement into progesterone

Progesterone is a molecule that cannot be directly radiolabeled. It is therefore necessary to couple it with another easily radiolabeled molecule, such as tyrosine methyl ester, histamine or gamma-globulin, prior to labeling (Figure 29).

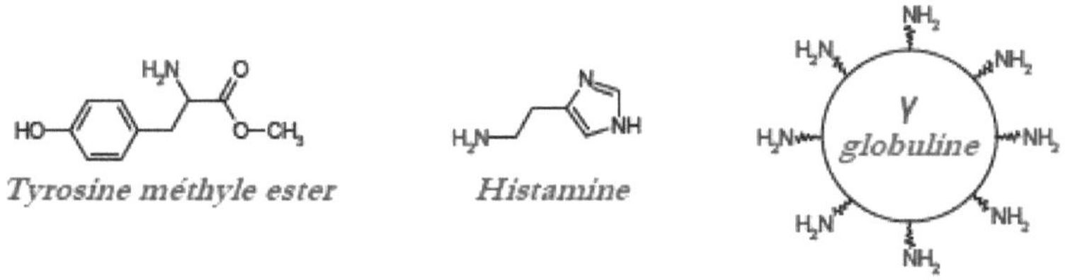

Figure 29: *Coupling molecules capable of being labelled at 125I*

Coupling will be performed with locally synthesized progesterone derivatives (P11α-HS or P3-CMO) using the "ester-activated" method or the "anhydride-mixed" method with tributylamine and isobutyl chloroformate as reagents (Figure 30).

We chose to use gamma-globulin as the coupling molecule for several reasons: it's easy to obtain, less expensive, easy to handle, and its coupling protocol is relatively simpler and faster (ester-activated method).

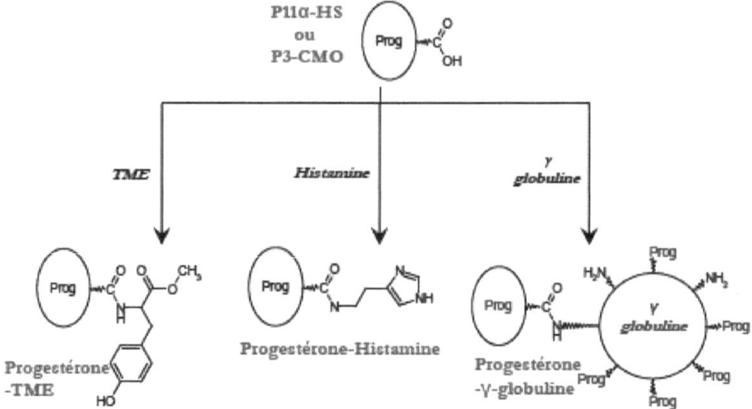

Figure 30: *Experimental schemes for coupling reactions*

II - COUPLING PROGESTERONE WITH GAMMA-GLOBULIN

Several chemical methods have been proposed for coupling antigens to gamma-globulin (Erlanger *et al.*, 1958; O'Rorke *et al.*, 1994; Basu *et al.*, 2003). The best are those that preserve the antigen binding site.

These methods are often divided into two categories: (i) direct covalent coupling, achieved by direct interaction or using a bifunctional reagent that establishes a covalent bridge between the two entities, and (ii) indirect coupling, involving two molecules with affinity for each other, each attached by a covalent bridge to the antigen on the one hand, and to the protein on the other (as in the case of the avidin - biotin system).

In our case, and in order to prepare radioactive progesterone tracers, we opted for direct covalent coupling between the hormone and the protein. For this, we used the "ester-activated" method, following a protocol similar to that we adopted for the preparation of immunogens (see Part 1, Paragraph I-2).

First, we prepared ester-activated derivatives of progesterone, then coupled these ester-activated derivatives with gamma-globulin.

1. Preparation of ester-activated derivatives

The ester-activated progesterone derivatives were prepared according to the same protocol described above for the preparation of immunogens. Thus, each of the two locally synthesized carboxyl-functional progesterone derivatives (P11α-HS or P3-CMO) enabled us to prepare an ester-activated derivative of this hormone.

The experimental protocol followed consists in dissolving 12 mg (~ 0.031 mmol) of the carboxyl-functional progesterone derivative (P11α-HS or P3-CMO) in a solution composed of 400 µl dimethylformamide (DMF) and 400 µl 1,4-dioxane. Next, 200 µl of a freshly prepared distilled water solution containing 12 mg of *N-hydroxysuccinimide* (NHS) and 24 mg of 1-ethyl-3-(3-dimethyl aminopropyl) carbodiimide hydrochloride (EDAC-HCl) is added. After homogenization, the reaction mixture was left at room temperature with gentle stirring overnight. Centrifugation at 15,000× *g* for 5 min then removes the urea acyl precipitate formed, and clarifies the solution containing the ester-activated derivative prepared.

2. Coupling with gamma-globulin

As in BSA coupling, each ester-activated progesterone derivative in solution is gently added to 4 ml of a solution containing 10 mg (0.250 µmol) of gamma-globulin in sodium borate buffer (0.1 M, pH 8) and kept at 4°C in an ice bath with gentle agitation. The reaction mixture is then incubated overnight at 4°C. The resulting progesterone-gamma globulin conjugate is dialyzed overnight at 4°C against 2 × 5 l phosphate buffer (0.010 M; pH 7.4) supplemented with 0.02% sodium azide (NaN$_3$). The volume of the dialyzed solution is then adjusted to 10 ml using the same dialysis buffer. This solution becomes the stock solution of the progesterone conjugate.

Eventually, two radioactive progesterone tracers were prepared. The difference between them lies in the binding position of progesterone with gmma-globulin (3 or 11). These two tracers (conjugated at position 3 or 11) will enable us to compare two progesterone immunoassay systems: homologous and heterologous, using antibodies produced from the two immunogens (conjugated at position 3 or 11).

III - PREPARATION OF RADIOACTIVE TRACERS

1. Incorporation routes for the 125I radioelement

Two major routes are used to incorporate the 125I radioelement:

- Incorporation of 125I into the protein by direct electrophilic substitution.

- Attachment of 125I to a chemical reagent which is then grafted onto the molecules to be radiolabeled.

In both cases, iodine, in its reduced form (NaI), reacts with the phenol group of a tyrosine or with the side chain of a histidine residue. These groups are first oxidized using an oxidizing agent.

The various oxidizing agents used :

- Chloramine T: strong oxidizer.
- Iodogen: a mild oxidizer.
- Lactoperoxidase: an enzyme that enables tyrosines to be oxidized in a controlled manner.

2. Marking reagents

The amount of radioelement incorporated differs according to the type of assay and the percentage of binding targeted. In the case of progesterone, where the analayte is found on serum, the labeling methodology is as follows:

- A: 25 µg of anti-TSH antibody
- B: 0.5 mCi (18.5 mBq) of 125I in 10 µl of 0.25 M phosphate buffer, pH 7.4
- C: 2.5 µg chloramine T in 10 µl 0.25 M phosphate buffer, pH 7.4
- D: 2.5 µg metabisulfute in 10 µl 0.25 M phosphate buffer, pH 7.4

3. Procedure

Solutions A, B and C are mixed in sequence. After 90 seconds, solution D is added, followed by 200 µl of 0.05 M phosphate buffer, pH 7.4 containing 2% BSA.

This mixture is transferred to a PD10 column (equilibrated with 0.05 M phosphate buffer, pH 7.4 containing 2% BSA for 30 min). Elution is carried out with the same buffer at 20-30 ml/h. Fractions of 0.5 ml are recovered and those containing radioactivity are pooled.

The column can be reused by washing for 2 h with distilled water containing 0.1% sodium azide.

4. Result

In this section, we prepared and evaluated the tracers, in order to select the conjugate (P11HS-GLOB or P3CMO-GLOB) that gives the best results for the locally prepared kit.

4-1. Purification of I125-labelled tracers

After tracer labelling and purification on the PD10 column (see previous paragraph), the elution profiles corresponding to each tracer are plotted by measuring the rate of radioactivity in coups/6S, placing the tube 5 cm from the gamma counter. Figure (31) shows the elution profiles of the 2 tracers (P11HS-GLOB and P3CMO-GLOB).

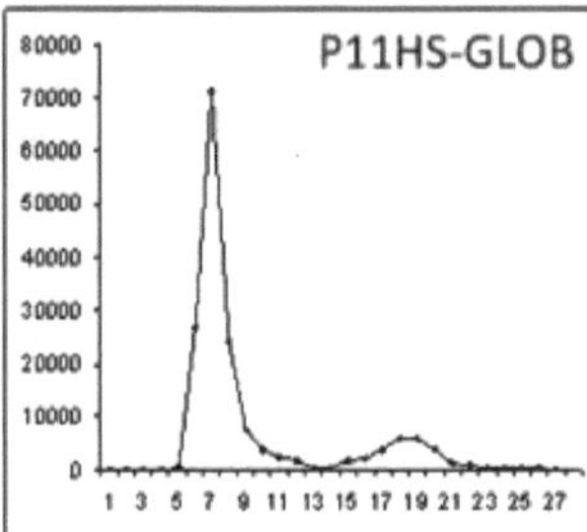

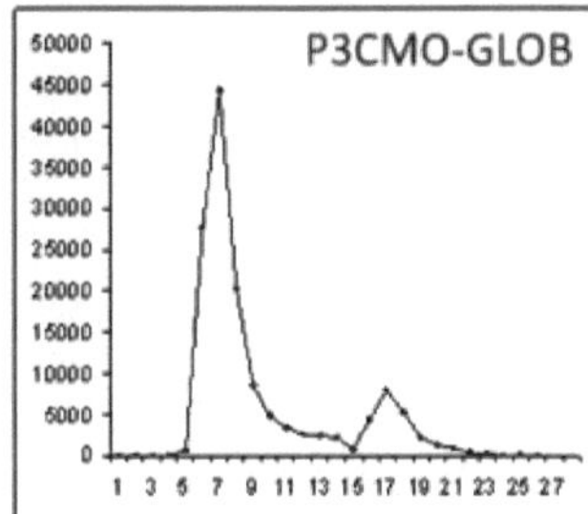

Figure 31: *Radiochromatogram of radioiodinated tracer elution*

These profiles reveal the presence of two peaks, one corresponding to tracer and the other to free iodide. Tracers are eluted first in the dead volume of the column, while free iodide is eluted later.

4-2. Marking yields

Marking yields are calculated for each tracer:

$$Rdt = \frac{Radioactivité - Traceur}{Radioactivité - Totale} \times 100$$

The labeling yields of the tracers are shown in Table 15 :

Table 15: *Iodine-125-labelled progesterone tracer labelling yields*

Plotters	Marking efficiency (%)
P11αHS-GLOB	83
P3CMO-GLOB	85

P11aHS-HRP : progesterone 11α-hemisuccinate - gamma globulin
P3CMO HRP : progesterone 3-(O-caroboxymethyl) oxime - gamma globulin

For the 2 tracers marked, we obtained good yields in excess of 80%.

In order to obtain the purest tracers possible, only 2 fractions with the highest level of radioactivity are collected. These fractions are then calibrated and tested for immunoreactivity against coated antibodies in serum.

IV - CALIBRATION OF RADIOACTIVE TRACER SOLUTIONS

1. Experimental protocol

- From the stock solution of each tracer, we prepared a series of 6 1:10 cascade dilutions in phosphate buffer (0.010 M; pH 7.4) supplemented with 0.02% sodium azide (NaN_3).

- 200 µl of each dilution is incubated for 3 h at room temperature with gentle agitation in a tube coated with anti-progesterone antibody (Diasource).

- To measure total radioactivity (T), the same volume (200 µl) of each dilution was incubated in a polystyrene tube (Nunc) containing no antibody. These tubes are revealed directly without washing.

- The coated tubes are then aspirated and washed 3 times with phosphate buffer (0.010 M; pH 7.4).

- Readings are taken immediately afterwards using a gamma counter (RIA-STAR).

2. Results

The results obtained, in terms of strokes per minute (B = CPM), are then converted into binding percentages relative to the total radioactivity recorded for each dilution (B0/T). The values obtained are shown in Table 16.

Table 16: *Binding percentages (B0/T) at different dilutions of locally prepared progesterone radioactive tracers (P11αHS-GLOB and P3CMO-GLOB)*

	Tracer dilution					
	1/10	*1/100*	*1/1000*	*1/10000*	*1/100000*	*1/1000000*
P11αHS-GLOB	0,26	5,01	17,15	50,07	100	100
P3CMO-GLOB	0,14	2,89	12,63	48,98	97,12	100

P11aHS-HRP : progesterone 11α-hemisuccinate - gamma globulin
P3CMO-HRP : progesterone 3-(O-caroboxymethyl) oxime - gamma globulin

These results show that for the two locally prepared progesterone radioactive tracers (P11αHS-GLOB and P3CMO-GLOB), the appropriate dilution for use in a competitive radioimmunoassay system would be 1/10000. Indeed, it is this dilution that enabled us to achieve a total binding percentage (B0/T) of around 50%, for both tracers tested separately.

Solutions of radioactive progesterone tracers were adjusted to the appropriate dilution using phosphate buffer (0.010 M; pH 7.4) supplemented with 0.02% sodium azide (NaN_3). These solutions are stored in aliquots at -20°C until use.

PART 4: DEVELOPMENT OF THE PROGESTERONE RADIOIMMUNOASSAY SYSTEM

As with most small molecules, radioimmunoassay of progesterone is generally carried out using the "competition" or "indirect" method. The principle of this method is to place the antigen to be assayed (analyte) in competition with the labeled antigen (tracer) in a well-defined quantity. The concentrations of antibody and labeled antigen are fixed. The higher the concentration of the antigen to be measured, the lower the quantity of labelled antigen fixed. The resulting calibration curve then decreases (Figure 29).

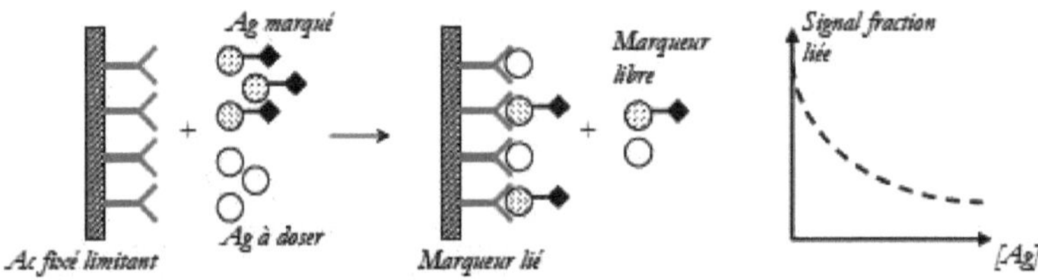

Figure 29: *Schematic diagram describing the principle of a competitive immunoassay*
Ac: antibody; Ag: antigen

The development of such a radioimmunoassay system (by competition) is classically carried out in two stages (Neuburger, 2006):

- First, a dilution range of the adsorbed antibody is performed in the presence of the tracer, in order to select the right antibody concentration for the assay. This concentration must generate a sufficient signal to perform the assay, while at the same time being limiting in order to favor competition from the analyte.

- Next, a calibration curve is performed in the presence of the analyte at different concentrations and the antibody at the chosen concentration. The concentrations of the samples to be assayed are then estimated in relation to this calibration curve.

These various steps will be more or less covered in this part of our work, in which we describe in greater detail the development of a radioimmunoassay system for progesterone specific to goats. This system, developed locally, is a heterogeneous format assay based on the use of an iodine-125-labeled tracer responsible for the appearance of a quantifiable signal.

I - PREPARATION OF TUBES COATED WITH ANTI-PROGESTERONE ANTIBODIES

1. *Coating*

This involves the attachment of active proteins (antibodies or other types) to solid supports. This process has been constantly used by biological diagnostic laboratories to develop immunoassay techniques (Couturier *et al.*, 1986). Several solid supports have been proposed for this purpose (polystyrene, nylon, cellulose, sepharose, polyacrylamide). Of these, polystyrene is the most popular.

Antibodies can be attached to the solid support in various ways:

> *Physical adsorption:* This takes place by simple contact of antibodies in a buffer solution with the solid support. The physical phenomenon that governs this adsorption is poorly understood, and antibody grafting is carried out in a rather empirical way.

> *Covalent bonding :* Binding is generally via a bifunctional reagent (glutaraldehyde, etc.). This reagent is able to bind specifically to the functional groups (COOH, NH_2 ...) of both the antibody and the solid support.

> *Coupling of antibodies to a pre-fixed protein:* Depending on the antibody's molecular weight, binding can be achieved with or without the intervention of a linker arm. The aim is to move the antibody away from the support wall to avoid steric hindrance and preserve antibody binding sites.

In our case, and in order to prepare tubes coated with polyclonal anti-progesterone antibodies (IgG or IgY), we opted for physical adsorption as the fixation method. Our choice is justified by the simplicity of this widespread technique, which has long been described as a gentle and effective method of fixation.

2. Materials required

Reagents	Special equipment

- NaH$_2$ PO$_4$ - 2H O$_2$	- Tubes (MaxiSorp, Nunc)
- Na$_2$ HPO$_4$ - 2H O$_2$	- Balance
- Sodium azide (NaN)$_3$	- pH meter
- Tris-HCl	- Vacuum system
- Citric acid	- Lyophilization
- Sodium citrate	- ELISA reader
- Casein	
- Sucrose	
- Distilled water	
- Anti-progesterone antibodies (IgG or IgY)	
- Radioactive progesterone tracers	

3. Experimental protocol

- From the stock solution of each anti-progesterone antibody, produced and purified as described previously (Part 1, Experimental study), we prepared a series of 5 1:10 cascade dilutions in buffer A (adsorption buffer) with the following composition: 0.315 g/l NaH$_2$ PO$_4$ - 2H$_2$ O; 1.2 g/l Na$_2$ HPO$_4$ - 2H$_2$ O and 0.5 g/l NaN$_3$ (pH 7.4).

- 200 µl of each dilution is incubated for 24 h at room temperature in a tube. Tubes (MaxiSorp, Nunc) are most often used, enabling a large number of samples to be processed simultaneously.

- The solutions contained in the tubes are then removed by aspiration and replaced with 300 µl/tube of buffer B (blocking buffer), composed as follows: 6 g/l Tris-HCl; 3.2 g/l citric acid; 10 g/l sodium citrate; 1 g/l NaN$_3$; 10 g/l casein and 50 g/l sucrose (pH 7.4). Tubes are then incubated for 24 h at room temperature.

- The tubes thus coated with anti-progesterone antibodies at different concentrations are aspirated, then dried for 30 min at 37°C under vacuum in a lyophilizer.

- Revelation can then be carried out using the radioactive progesterone tracer (prepared locally) at a rate of 200 µl/tube, following the usual revelation protocol. The aim is to determine, for each anti-progesterone antibody, the dilution that will generate sufficient signal to perform the assay, while being limiting in order to favor competition (B0/T ~ 50%).

4. Result

For each dilution, the results obtained in terms of strokes per minute (B = CPM) are converted into binding percentages relative to the total activity recorded by the radioactive tracer (B0/T). The values obtained are shown in Table 17.

Table 17: *Binding percentages (B0/T) at different antibody dilutions polyclonal anti-progesterone (IgG and IgY)*

	Antibody dilution				
	1/100	*1/1000*	*1/10000*	*1/100000*	*1/1000000*
Anti-progesterone IgG	99,87	47,38	15,01	2,45	0,14
IgY anti-progesterone	99,25	46,76	12,91	3,00	0,00

These results show that for the two anti-progesterone antibodies (IgG and IgY) used to prepare the coated tubes, the appropriate dilution for use in a competitive radioimmunoassay system is 1/1000. This dilution enabled us to achieve a total binding percentage (B0/T) of between 40 and 50% (Table 16).

From now on, all tubes coated with anti-progesterone antibodies, to be used in our radioimmunoassay system for progesterone, will be prepared (by *coating*) from 1/1000 dilutions of these antibodies.

II - SETTING UP THE PROGESTERONE RADIOIMMUNOASSAY SYSTEM

After several preliminary trials, we opted for the so-called "heterologous format" setup for our radioimmunoassay of progesterone. This choice is supported by the numerous arguments cited in the literature in favor of this type of setup for steroid assays (Mitsuma *et al.*, 1987; Basu *et al.*, 2006). Consequently, the anti-progesterone antibodies to be used in our assay system are those produced against the immunogen conjugated to position 11 of progesterone (P11α-HS - BSA), while the radioactive tracer to be employed is that conjugated to position 3 of progesterone (P3-CMO - I125).

1. Components of the progesterone radioimmunoassay system

The various components of our radioimmunoassay system, developed locally to measure progesterone in goats, are now ready for use in control and validation trials (Figure 30).

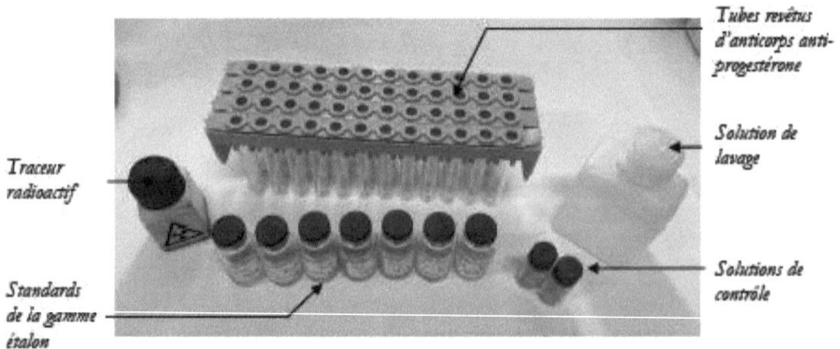

Figure 30: *Components of the progesterone radioimmunoassay kit specific to goats, developed locally*

2. Determination of progesterone and establishment of the calibration curve

2-1. Experimental protocol

From each local progesterone standard, 50 µl is added to a tube coated with anti-progesterone antibody. 200 µl of the solution containing the radioactive progesterone tracer at the appropriate concentration is then added to each tube.

After incubation for 3 h at room temperature with 100 rpm agitation, tubes are aspirated, then washed with a washing solution composed of: 20 mM Tris-HCl; 150 mM NaCl; 0.05% (v/v) tween 20 (pH 7.5). This operation is repeated twice to completely remove unbound tracer. Readings are taken immediately afterwards using a RIA-STAR gamma counter.

2-2. Results

The results obtained, in terms of strokes per minute (B = CPM), are then converted into binding percentages relative to the maximum enzyme activity recorded in the absence of free progesterone (B/B0). The values obtained are shown in Table 18.

Table 18: *Percentage of bonding (B/B0) corresponding to each standard of our local progesterone standard range*
Measurements made using our own progesterone enzyme immunoassay system

Standard	[Progesterone] (ng/ml)	CPM	B/B0× 100 (%)
S_0	0, 000	1815	100

S_1	0, 374	1606	88, 48
S_2	0, 874	1351	74, 43
S_3	1, 829	1029	56, 69
S_4	5, 687	0523	28, 81
S_5	15, 14	0232	12, 78
S_6	29, 66	0115	6, 33

By plotting the percentage binding (B/B0 × 100) calculated for each standard against the logarithm of the corresponding progesterone concentration, we obtain the classic representation of the calibration curve for our progesterone radioimmunoassay system (Figure 31 A). Another representation of this calibration curve can also be established. This so-called "Logit-Log" representation consists in plotting "Log (B/B0-B) = f (Log [Progesterone])". The resulting calibration curve turns out to be a straight line, from which it is easier to extrapolate (Figure 31 B).

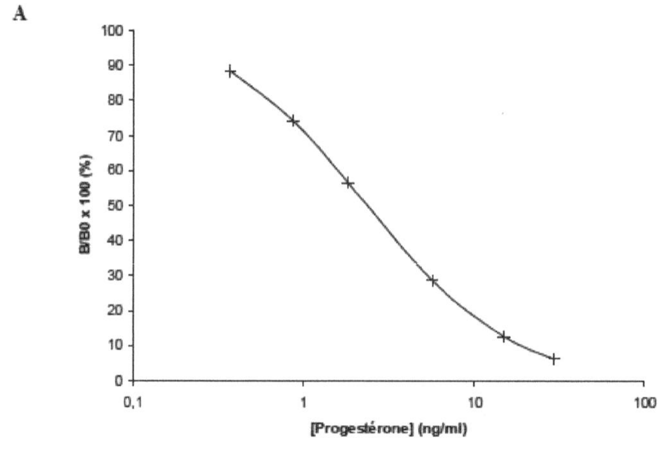

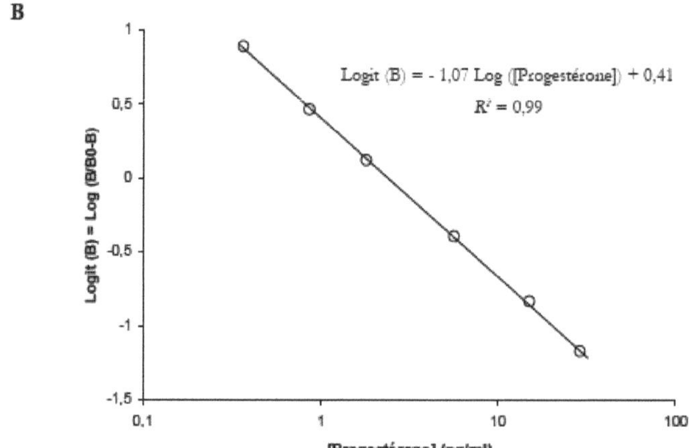

Figure 31: *Classical (A) and linear (B) representations of the calibration curve for the locally developed progesterone radioimmunoassay system.*

In both conventional and linear representations (Figure 31 A and B), the calibration curve for our locally-developed radioimmunoassay system for progesterone shows a consistent profile, with a high degree of proportionality between progesterone concentration and the signal recorded by the ELISA reader after revelation. The resulting regression line (Figure 31 B) reveals a correlation coefficient of 0.99.

This result confirms the conformity and high quality of the radioimmunoassay system for progesterone, which we have developed in our laboratory to measure this hormone in goats. Studies are currently underway to validate the methodological and statistical aspects of this assay system.

We also plan to carry out other types of set-up using other compounds, prepared locally, in order to compare them and choose from them the "antibody - tracer" pair best suited to the radioimmunoassay of progesterone.

CONCLUSION

In socio-economic terms, goats make an effective contribution to generating income and meeting the milk and meat requirements of a large segment of the Moroccan rural population in most of the kingdom's remote and/or isolated areas (Fares and Ghalim, 1982; Jout and Karimi, 2004). Aware of this vital interest, the public authorities have set up a goat breeding development program aimed, among other things, at improving the production potential of the flocks. In order to achieve this objective, better control of reproduction is an absolute must.

Among the biotechnological methods that can be explored in this respect is the measurement of reproductive hormones, and in particular progesterone. Estimating plasma or peripheral serum progesterone levels is a relevant experimental tool, widely used throughout the world to determine the physiological state of females on a farm.

This is where the interest lies in the present study, whose main objective is to develop a radioimmunoassay system for progesterone, with a view to its application in the management of goat reproduction. The aim of this project is to provide breeders and livestock technicians with a valuable tool enabling them to improve and better organize the management of their flocks.

To develop this progesterone immunoassay system, we first set out to locally produce highly specific polyclonal anti-progesterone antibodies. This was made possible by the use of certain chemical coupling techniques, which enabled us to prepare the necessary immunogens. These were obtained by grafting progesterone onto a carrier protein (BSA) at certain positions on the hormone's steroid nucleus.

As progesterone is a molecule that doesn't lend itself directly to this kind of reaction, we were obliged to synthesize two of its carboxyl-functional derivatives that could be more easily bound to the carrier protein. These were progesterone 11α-hemisuccinate (P11α-HS) and progesterone 3-(O-caroboxymethyl) oxime (P3-CMO). The synthetic yields of these two compounds were estimated at 70 and 74% respectively. A comprehensive structural study, using standard analytical chemistry techniques (TLC, IR spectrometry, NMR[1] H, mass spectrometry), enabled us to characterize these two locally synthesized derivatives.

Coupling of these derivatives (P11α-HS and P3-CMO) to BSA was then carried out using the "ester-activated" method, which involves preparing an ester-activated derivative of the steroid from the carboxyl-functional derivative, and then coupling this ester-activated derivative to the protein. In this way, we were able to prepare two progesterone immunogens, conjugated to one of the two most preferred positions of the steroid nucleus (3 and 11). The coupling yields of these two immunogens were found to exceed 20 progesterone molecules per BSA molecule, as estimated by spectrophotometric analysis and supported by electrophoretic analysis. We can therefore say that, in addition to their susceptibility to being engaged in chemical coupling reactions, the two locally synthesized progesterone derivatives (P11α-HS and P3-CMO) also offer the possibility of directing these reactions, thus enabling the preparation of more appropriate immunogens.

Thanks to these immunogens, we were able to produce two types of polyclonal anti-progesterone antibodies (IgG and IgY) in two different animal species (rabbit and hen), with very high affinities and specificities - prerequisites for the development of a good immunoassay system. These polyclonal antibodies, produced locally, were purified using various biochemical separation techniques (fractional salt precipitation, chromatography, etc.).

In order to prepare the radioactive progesterone tracers needed to develop our assay system, we labelled locally synthesized carboxyl-function progesterone derivatives with iodine-125. The labelling protocol used was that of chloramine T. The radioactive tracers thus obtained, conjugated to one of the two most recommended positions of the steroid nucleus (3 and 11), can then be used, together with locally produced anti-progesterone antibodies, to set up different types of progesterone radioimmunoassay system (homologous or heterologous).

As any specific immunoassay procedure is based on an appropriate reference system, it was essential for us to consider preparing our own progesterone standard range locally. For this purpose, we opted for depleted goat serum as the assay matrix, to better reproduce the conditions under which our assay system will be used. This matrix was overloaded with pure crystalline progesterone at well-defined concentrations in order to prepare the different standards making up our standard range. Once prepared, the local progesterone standard was statistically validated according to international standards (Caporal-Gautier *et al.*, 1992; Albrecht *et al.*, 2004) against a number of validation criteria.

The various components of our enzymo-immunoassay system, developed locally to measure progesterone in goats (anti-progesterone antibodies, radioactive tracer and standard range), were then used in a device in heterologous format to carry out assay tests. Calibration of the solution containing the radioactive tracer and adjustment of the concentration of the anti-progesterone antibodies used to prepare the coated tubes were essential to reproduce competition conditions. All the results obtained were judged to be analytically satisfactory, enabling us to conclude that the RIA kit, produced locally for the determination of progesterone in goats, has all the qualities expected of a good radioimmunoassay kit.

These encouraging results have prompted us to initiate the procedures for methodological validation of our assay system, in order to further assess its various analytical performances (detection limit, specificity, precision, accuracy). Clinical validation by goat breeding specialists will then complete the final steps necessary to qualify our assay.

In line with the objectives of this study, we plan to prepare enzymatic tracers for progesterone, which will be used to develop an enzymo-immunoassay system for this hormone. This will provide an alternative to the radioimmunoassay system already developed.

The development of other immunoassay systems, in line with the overall strategic framework outlined by this project, is also envisaged in the long term. These include immunoassays for testosterone, LH and PAG. Together with progesterone, these assays are excellent tools that can be used to study the mechanisms of goat reproductive physiology, providing the basis for better organization of reproductive management.

REFERENCES

Agasan A.L., Stewart B.J. & Watson T.G. (1994). Development of a radioimmunoassay method for ethynylestradiol in plasma using a monoclonal antibody. *J. Immunol. Methods,* vol. **177**, p. 251-260.

Albrecht B., Balsiger C., Bogli R., Buxtorf U.P., Buhler F., Emch H., Jacob A., Gremaud G., Hubner P., Luginbuhl W., Roos P., Schurter M., Spack I., Wenk P. & Wolfensberger M. (2004). Guide to the validation of physico-chemical test methods and the assessment of measurement uncertainty. *Swiss Food Manual*, p. 1-27.

Allen R.M. & Redshaw M.B. (1978). The use of homologous and heterologous[125] I-radioligands in the radioimmunoassay of progesterone. *Steroids*, vol. **32**, p. 467-486.

Argemi B. (1998). Indications for steroid hormone assays. *Revue de l'ACOMEN*, vol. **4** (3), p. 225-231.

Bacigalupo M.A., Ferrara L., Meroni G. & Ius A. (1987). Time-resolved fluoroimmunoassay of progesterone by Eu-libelled protein-A. *Fresenius Z. Anal. Chem.* vol. **328**, p. 263-264.

Baril G., Brebion P. & Chesné P. (1993). Practical training manual for embryo transplantation in ewes and goats. *FAO "Production et Santé Animales"*, Rome, Italy, 115 p.

Barkley M.S., Lasley B.L., Thompson M.A. & Shackleton C.H. (1985). Equol: a contributor to enigmatic immunoassay measurements of estrogen. *Steroids*, vol. **46**, p. 587-608.

Basu A., Shrivastav T.G. & Kariya K.P. (2003). Preparations of enzyme conjugate through adipic acid dihydrazide as linker and its use in immunoassays. *Clin. Chem.* vol. **49**, p. 1410-1412.

Basu A., Shrivastav T.G. & Maitra S.K. (2006). A direct antigen heterologous enzyme immunoassay for measuring progesterone in serum without using displacer. *Steroids*, vol. **71**, p. 222-230.

Baulieu E.E. (1992). Mechanism of action of steroid hormones and anti-hormones. A 1991 mini-overview. *C. R. Acad. Sci. III*, vol. **314**, p. 23-26

Benlakhal A. (2004). Introduction. In: Chriqi A. (Ed). *Elevage caprin : Quelle stratégie de développement. 7ème édition de la foire caprine de Chefchaouen*. Chefchaouen, Morocco, p. 11-12.

Billings H.J. & Katz L.S. (1997). Progesterone facilitation and inhibition of estradiol-induced sexual behavior in the female goat. *Hormones and Behaviour*, vol. **31**, p. 47-53.

Billings H.J. & Katz L.S. (1998). Threshold dose of estradiol for inducing sexual receptivity in ovariectomized French Alpine goats. *Appl. Anim. Behav. Sci.* vol. **57**, pp. 109-115.

Brice G. (2003). Luminous deseasoning in goat production. Published by Institut de l'Elevage in October 2003, Paris, France, 40 p.

Brossier P., Jaouen G., Limoges B., Salmain M., Vessieres-Jaouen A. & Yvert J.P. (2001). Contribution of simultaneous multiple immunoassays to biology. *Ann. Biol. Clin (Paris)*, vol. **59** (6), p. 677-691.

Caporal-Gautier J., Nivet J.M., Algranti P., Guilloteau M., Histe M., Lallier M., N'Guyen-Huu J.J. & Russotto R. (1992). Guide de validation analytique. *Rapport d'une commission SFSTP (I/ Méthodologie & II/ Exemples d'application), STP PharmaPratiques*, vol **4**, p. 205-226.

Catt K., Niall H.D. & Tregear G.W. (1967). Solid phase radioimmunoassay. *Nature*, vol. **213** (78), p. 825-827.

EEC III/844/87-EN. (1989). Final explanatory note

Champigny G., Voilley N., Lingueglia E., Friend V., Barbry P. & Lazdunski M. (1994). Regulation of expression of the lung amiloride-sensitive Na^+ channel by steroid hormones. *EMBO J.,* vol. **13**, p. 2177-81

Chandrasekhar T. & Madan M.L. (1996). Active immunization againt estrone in goats and its effects on ovarian hormones and cyclicity. *Indian J. Anim. Sci.* vol. **66**, pp. 990-993.

Chemineau P., Gauthier D., Poirier J.C. & Saumande J. (1982). Plasma levels of LH, FSH, prolactin, estradiol-17 beta and progesterone during natural and induced estrus in the dairy goat. *Theriogenology*, vol. **17**, p. 313-323.

Chemineau P., Malpaux B., Guérin Y., Maurice F., Daveau A. & Pelletier J. (1992). Light and melatonin for the control of reproduction in sheep and goats. *Ann. Zootech.* vol. **41**, 247-261.

Chentouf M. (2007). Reproductive physiology and productivity of the northern Moroccan goat. Doctoral thesis in Veterinary Sciences. *Faculté des Sciences, Université Notre-Dame de la Paix*, Namur, Belgium, 161 p.

Chentouf M., Ayadi M. & Boulanaouar B. (2004). Typologie des élevages caprins dans la province de Chefchaouen : Fonctionnement actuel et perspectives. *Options Méditerranéennes*, vol. **61**, p. 255-261.

Christofidis I., Noikokyri-Kouvalaki E., Mastichiadis C., Petrou P.S. & Kakabakos S.E. (2006). Development of radioimmunoassay system for the determination of progesterone in cow milk. In: Proceedings of a final research coordination meeting held in Vienna, 6-10 December 2004 *"Development of radioimmunometric assays and kits for non-clinical applications"*, IAEA-TECDOC-1498, p. 49-65.

Cousino M.A., Jarbawi T.B., Halsall H.B. & Heineman W.R. (1997). Pushing down the limits of detection: molecular needles in a haystack. *Anal. Chem.* vol. **69** (17), p. 544A-549A.

Couturier R., Ville A., Perrin B. & Favre-Bonvin G. (1986). Covalent binding of antibodies to polystyrene. Activation of the support and repeated use of the polystyrene-antibody system after enzyme immunoassay. *Macromolecular Chemestry and Physics*, vol. **187** (7), p. 1603-1610.

Creminon C., Dery O., Frobert Y., Couraud J.Y., Pradelles P. & Grassi J. (1995). Two-Site Immunometric Assay for Substance P with Increased Sensitivity and Specificity. *Anal. Chem.* **67** (9), pp. 1617-1622.

Darwati S., Ariyanto A., Yunita F., Mondrida G., Triningsih A., Setyowati S., Sutari A., Sovilawati E. & Martalena A. (2006). Development of radioimmunoassay kits for non clinical applications: local production of primary reagents for milk progesterone. In: Proceedings of a final research coordination meeting held in Vienna, 6-10 December 2004 *"Development of radioimmunometric assays and kits for non-clinical applications"*, IAEA-TECDOC-1498, p. 49-65.

Dmitriev D.A., Massino Y.S. & Segal O.L. (2003). Kinetic analysis of interactions between bispecific monoclonal antibodies and immobilized antigens using a resonant mirror biosensor. *J. Immunol. Methods*, vol. **280** (1-2), p. 183-202.

Dutruc-Rosset G. (1999). Protocol for validating a conventional analytical method against the O.I.V. reference method.

Edelman G.M., Cunningham B.A., Gall W.E., Gottlieb P.D., Rutishauser U. & Waxdal M.J. (1969). The covalent structure of an entire gamma G immunoglobulin molecule. *Proc. Natl. Acad. Sci. U. S. A*, vol. **63** (1), p. 78-85.

Ekins R.P. (1993). New perspectives in radioimmunoassay. *Nucl. Med. Commun.* vol. **14** (9), p. 721-735.

El Amiri B., Karen A., Cognie Y., Sousa N.M., Hornick J.L., Szenci O. & Beckers J.F. (2003). Diagnosis and monitoring of pregnancy in ewes: realities and perspectives. *INRA Prod. Anim*, vol. **16** (2), p. 79-90.

El Fadil H. (1994). Dairy performance of the local goat: Quantitative and qualitative study. *Doctoral thesis in Veterinary Medicine*. IAV Hassan II, Rabat, Morocco.

Engvall E. & Perlman P. (1971) Enzyme-linked immunosorbent assay (ELISA). Quantitative assay of immunoglobulin G. *Immunochemistry*, vol. **8** (9), pp. 871-874.

Erlanger B.R., Borek F., Beiser S.M. & Lieberman S. (1958). Steroid-protein conjugates, II - Preparation and characterization of conjugates of bovine serum albumin with progesterone, deoxycorticosterone, and estrone. *The Journal of Biological Chemistry*, vol. **234**, p. 1090-1094.

Fabre H. (1999). Validation of capillary electrophoresis methods applied to the analysis of pharmaceutical compounds. *Analusis*, vol. **27**, p. 155-160.

Fabre-Nys C. (2000). Sexual behavior in goats: hormonal control and social factors. *INRA Prod. Anim.* vol. **13** (1), p. 11-23.

Fabre-Nys C., Martin G.B. & Venier G. (1993). Analysis of the hormonal control of female sexual behavior and the preovulatory LH surge in the ewe: Roles of quantity of estradiol and duration of its presence. *Hormones and Behaviour*, vol. **27**, p. 108-121.

Fares A. & Ghalim A. (1982). Goat breeding in the Haut Loukkos: production system and development prospects. *Mémoire de 3ème cycle en Agronomie*, Ecole Nationale d'Agriculture de Meknès. Meknès. Morocco.

Fivash M., Towler E.M. & Fisher R.J. (1998). Biacore for macromolecular interaction. *Curr. Opin. Biotechnol*, vol. **9** (1), pp. 97-101.

Freitas V.J.F., Baril G., Martin G.B. & Saumande J. (1997). Physiological limits to further improvement in the efficiency of oestrous synchronisation in goats. *Reprod. Fert. Develop.* vol. **9**, p. 551-556.

Gonzalez S.C., Corteel J.M. & Baril G. (1992). Cinetica de la progesterona plasmatica durante el celo natura e iducido por tratamientos hormonales en cabras lecheras. *Revistica cientifica, FCV de LUZ*, vol. **II** (1), p. 12-21.

Gordon I. (1997). Controlled reproduction in sheep and goats. *CAB International publ.*

Hachi A. (1990). La chèvre D'man : Contribution à l'étude des caractéristiques de reproduction. *Doctoral thesis in Veterinary Medicine.* IAV Hassan II, Rabat, Morocco.

Harma H., Soukka T., Lonnberg S., Paukkunen J., Tarkkinen P. & Lovgren T. (2000) Zeptomole detection sensitivity of prostate-specific antigen in a rapid microtitre plate assay using time-resolved fluorescence. *Luminescence*, vol. **15** (6), p. 351-355.

Hassani H. (1997). Impact de l'élevage des populations sur les ressources naturelles dans le Rif Occidental: Cas de la commune rurale de Béni Idder. *Mémoire de 3ème cycle en Agronomie*, Ecole Nationale d'Agriculture de Meknès, Meknès, Maroc.

Homeida A.M. & Cooke R.G. (1984). Plasma concentrations of testosterone and 5 alpha-dihydrotestosterone around luteolysis in goats and their behavioural effects after ovariectomy. *J. Steroid Biochem.* vol. **20**, p. 1357-1359.

Horie H., Kidowaki T., Koyama Y., Endo T., Homma K., Kambegawa A. & Aoki N. (2007). Specificity assessment of immunoassay kits for determination of urinary free cortisol concentrations. *Clin. Chim. Acta.* vol. **378** (1-2), p. 66-70.

Janin J. (1995). Principles of protein-protein recognition from structure to thermodynamics. *Biochimie*, vol. **77** (7-8), pp. 497-505.

Janoski A.H., Shulman F.C. & Wright G.E. (1973). Selective 3-(O-carboxymethyl) oxime formation in steroidal 3,20-diones for hapten immunospecificity. *Steroids*, vol. **23**, p. 49-64.

Jardy A. & Vial J. (1999). L'apport des méthodes statistiques dans la maîtrise de la qualité des analyses. *Analusis*, vol. **27**, p. 511-522.

Jout J. & Karimi A. (2004). Status and problems of goat development in the northern zone. In Chriqi A. (Ed). *Elevage caprin : Quelle stratégie de développement. 7ème édition de la foire caprine de Chefchaouen. Chefchaouen, Morocco*, pp. 13-20.

Kamoun P. (1997). Apparatus and methods in biochemistry and molecular biology. *Médecine - Science, Edition Flammarion*, Paris, France, 432 p.

Kaplan D.H. & Katz L.S. (1994). Exposure to constant photoperiod alters serum prolactin concentrations and behavioural response to estradiol in the ovariectomized goat. J. *Anim. Sci.* vol. **72**, pp. 3088-3097.

Karush F. (1958). Specificity of antibodies. *Trans. N. Y. Acad. Sci.* vol. **20** (7), p. 581-592.

Köhler G. & Milstein C. (1975). Continuous cultures of fused cells secreting antibody of predefined specificity. *Nature*, vol. **256** (5517), p. 495-497.

Kothari K, Lal R. & Pillai M. R. A. (1995). Development of a direct radioimmunoassay for serum progesterone. *Journal of Radioanalytical and Nuclear Chemistry*, vol. **196**, p. 331-338.

Kothari K. & Pillai M.R.A. (1998). Direct radioimmunoassay of serum progesterone using heterologous bridge tracer and antibody. *Journal of Radioanlytical and Nuclear Chemistry*, vol. **231**, p. 77-82.

Lemon M. & Thimonier J. (1973). Evolution of plasma progesterone during cycle and gestation in ruminants. In: Denamur R. & Netter A. (Eds), *Colloque Société Nationale pour l'Etude de la Stérilité et de la Fécondation : " corps jaune "*, p. 51-68, Masson, Paris.

Ling C.M. & Overby L.R. (1972). Prevalence of hepatitis B virus antigen as revealed by direct radioimmune assay with 125 I-antibody. *J. Immunol.* vol. **109** (4), p. 834-841.

Lupi-Chen N., Hoang B.M. & Cailla H. (1999). Immunoassays for steroids: testosterone, progesterone and cortisol in animals. *Immuno analyse & biologie spécialisée*, vol. **14** (4), p. 269-275.

MADRPM (2004). Situation de l'agriculture Marocaine. *Ministère de l'Agriculture, du Développement Rural et de la Pêche Maritime*, Rabat, Morocco, 104 p.

MAPM (2008). Situation de l'élevage au Maroc. *Ministère de l'Agriculture et de la Pêche Maritime*, Rabat, Morocco.

Mercier-Bodard C., Alfsen A. & Baulieu E.E. (1970). Sex steroid binding plasma protein (SBP). *Acta Endocrinol. suppl.* vol. **147**, p. 204-224.

Mickelson K.E., Forsthoefel J. & Westphal U. (1981). Steroid-protein interactions. Human corticosteroid binding globulin: some physicochemical properties and binding specificity. *Biochemistry,* vol. **20**, p. 6211-6218.

Miles L.E. & Hales C.N. (1968). Immunoradiometric assay of human growth hormone. *Lancet*, vol. **2** (7566), p. 492-493.

Mitsuma M., Kambegawa A., Okinaga S. & Arai K. (1987). A sensitive bridge heterologous enzyme immunoassay of progesterone using geometrical isomers. *J. Steroid Biochem.* vol. **28**, p. 83-88.

Mori Y. & Kano Y. (1984). Changes in plasma concentrations of LH, progesterone and estradiol in relation to the occurrence of luteolysis, estrus and time of ovulation in the Shiba goat (Capra hircus). *J. Reprod. Fert*, vol. **72**, p. 223-230.

Moulin M. & Coquerel A. (2002). Pharmacologie. *2ème edition*, Masson, Paris, France, 592 p.

Neuberger L.M. (2006). Development of fluorescence immunoassays: Perspectives for the development of a flow sensor of threat agents. *Doctoral thesis, Institut National Agronomique Paris-Grignon*, Paris, France, 336 p.

Niswender G.D. (1973). Influence of the site of conjugation on the specificity of antibodies to progesetrone. *Steroids*, vol. **22**, p. 413-424.

O'Rorke A., Kane M.M., Golsing J.P., Tallon D.F. & Fottrell P.F. (1994). Development and validation of a monoclonal antibody enzyme immunoassay for measuring progesterone in saliva. *Clin. Chem.* vol. **40** (3), p. 454-458.

Okada M., Hamada T., Takeuchi Y. & Mori Y. (1996). Timing of proceptive and receptive behavior of female goats in relation to the preovulatory LH surge. *J. Vet. Med. Sci.* **58**, pp. 1085-1089.

Okada M., Takeuchi Y. & Mori Y. (1998). Estradiol-dependency of sexual behavior manifestation at the post-LH surge period in ovariectomized goat. *J. Reprod. Develop.*, vol. **44**, p. 53-58.

Olivereau J.M. (1970). Hypothalamo-hypophyseal complex and regulation of water and electrolyte metabolism in the albino rat submitted to the influence of negative air ions. *Z. Zellforsch. Mikrosk. Anat.* vol. **105** (3), p. 430-441.

Pelizzola D., Bombardieri E., Brocchi A., Cappelli G., Coli A., Federghini M. *et al.* (1995). How alternative are immunoassay systems employing non-radioisotopic labels? A comparative appraisal of their main analytical characteristics. *Q. J Nucl. Med.* vol. **39** (4), p. 251-263.

Pocock G. & Christopher D.R. (2004). Human physiology. Edition Masson, Paris, France.

Poljak R.J., Amzel L.M., Avey H.P., Chen B.L., Phizackerley R.P. & Saul F. (1973). Three-dimensional structure of the Fab' fragment of a human immunoglobulin at 2,8-A resolution. *Proc. Natl. Acad. Sci. U. S. A*, vol. **70** (12), pp. 3305-3310.

Polson A., Von Wechmar M.B. & Van Regenmortel M.H.V. (1980). *Immunol. Commun.* vol. **9**, p. 475-493.

Porter R.R. (1967). The structure of immunoglobulins. *Biochem. essays*, vol. **3**, p. 1-24.

Pradelles P., Grassi J., Creminon C., Boutten B. & Mamas S. (1994). Immunometric assay of low molecular weight haptens containing primary amino groups. *Anal. Chem.* vol. **66**, p. 16-22.

Ruckebush Y., Phaneuf L.P. & Dunlop R. (1991). Physiology of small and large animals. Philadelphia, Hamilton, USA, 672 p.

Samuel G., Karir T., Kothari K., Joshi S., Sivaprasad N. & Venkatesh M. (2006). Development of radioimmunoassay for estimation of progesterone in bovine serum. In : Proceedings of a final research coordination meeting held in Vienna, 6-10 Decembre 2004 " *Development of radioimmunometric assays and kits for non-clinical applications* ", IAEA-TECDOC-1498, pp. 49-65.

Sawada T., Takahara Y. & Mori J., (1995). Secretion of progesterone during long and short days of the estrous cycle in goats that are continuous breeders. *Theriogenology*, vol. **43**, p. 789-795.

Schultz E., Perraut F., Neuburger L.M. & Volland H. (2003). Study of a highly sensitive fluorescence immunoassay system. *Colloque " Méthodes et Techniques optiques pour l'industrie " (Belfort, November 2003)*.

Simersky R., Swaczynova J., Morris D.A., Franek M. & Strnad M. (2007) Developpement of an ELISA-based kit for the on-farm determination of progesterone milk. *Veterinari Medicina*, vol. **52** (1), p. 19-28.

Simmer H.H. (1975). On the beginnings of hormonal contraception (author's translation). *Geburtshilfe Frauenheilkd*, vol. **35**, p. 688-696.

Sousa N.M., Gabrayo J.M., Figueiredo J.R., Sulon J., Gonçalvez P.B.D. & Beckers J.F. (1999). Pregnancy-associated glycoprotein and progesterone profiles during pregnancy and postpartum in native goat from the north-east of Brazil. *Small Rum. Res.* vol. **32**, p. 137-147.

Sousa N.M., Gonzalez F., Karen A., El Amiri B., Sulon J., Baril G., Cognie Y., Szenci O. & Beckers J.F. (2004). Diagnosis and monitoring of pregnancy in goats and ewes. *Renc. Rech. Ruminants*, vol. **11**, p. 377-380.

Sutherland S.R. & Lindsay D.R. (1991). Ovariectomized does not require progesterone priming for oestrous behaviour. *Reprod. Fert. Develop.* **3**, p. 679-684.

Thimonier J. (2000). Determination of the physiological state of females by analysis of progesterone levels. *INRA Prod. Anim*, vol. **13** (3), p. 177-183.

Valancia J., Zarco L., Ducoing A., Murcia D., Navarro H. (1990). Breeding season of Criollo and Granadina goats under constant nutritional level in the Mexican highlands, in *Proceedings of the Final Research Co-ordination Meeting*. International Atomic Energy Agency, Vienna, pp. 321-333.

Van Regenmortel M.H. (1994). In: *The Recognition of Proteins and Peptides by Antibodies*, pp. 277-300.

Van Regenmortel M.H. (1998). Mimotopes, continuous paratopes and hydropathic complementarity: novel approximations in the description of immunochemical specificity. *J. Dispersion Science and Technology*, vol. **19** (6&7), p. 1199-1219.

Vanroose G., de Kruif A. & Van Soom A. (2000). Embryonic mortality and embryo-pathogen interactions. *Animal Reproduction Science*, vol. 60-61, p. 131-143.

Volland H. (1999). *Application du procédé SPIE-IA au dosage du Leucotriène C4, de l'Angiotensine II et de l'Histamine*. Biologie Cellulaire et Moléculaire, PhD thesis, Université Pierre et Marie Curie - Paris VI, 266 p.

Wide L., Bennich H. & Johansson S.G. (1967). Diagnosis of allergy by an in-vitro test for allergen antibodies. *Lancet*, vol. **2** (7526), p. 1105-1107.

Wudt S.A., Wachter U.A., Homoki J. & Teller W.M. (1995). 17 alpha-hydroxyprogesterone, 4-androstenedione, and testosterone profiled by routine stable isotope dilution/gas chromatography-mass spectrometry in plasma of children. *Pediatr. Res.* vol. **38**, p. 76-80.

Yalow R.S. & Berson S.A. (1960). Immunoassay of endogenous plasma insulin in man. *Journal of Clinical Investigation*, vol. **39**, p. 1157-1175.

Yatsimirskaya E. A., Gavrilova E.M., Egorov A.M. & Levashov A.Y. (1993). Preparation of conjugates of progesterone with bovine serum albumin in the reversed micellar medium. Steroids, vol. 58, p. 547-550.

Zerr-Fouineau M. (2006). Characterization of the prothrombotic effect of progestogens at the level of the vascular wall: role of the endothelium and influences of estrogens. Doctoral thesis, Université Louis Pasteur de Strasbourg, Strasbourg, France, 197 p.

Zuber E. (1997) *Approche biométrique de la dynamique de réactions antigène-anticorps, cas de deux systèmes automatisés de suivi en temps réel des cinétiques*. PhD thesis: Biochemistry, Université Claude Bernard Lyon I, France, 223 p.

Printed by Books on Demand GmbH, Norderstedt / Germany